T3-BOJ-943

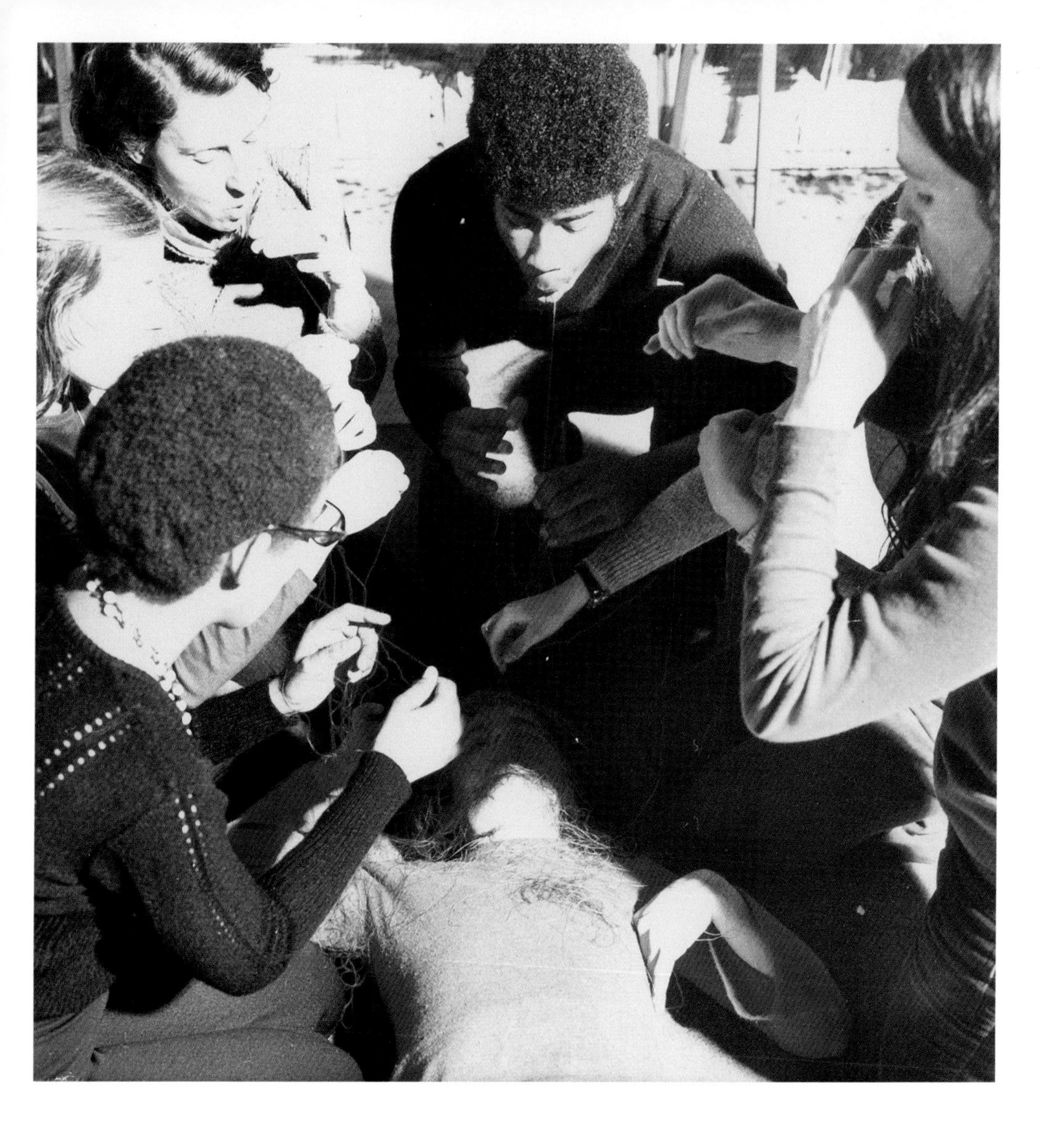

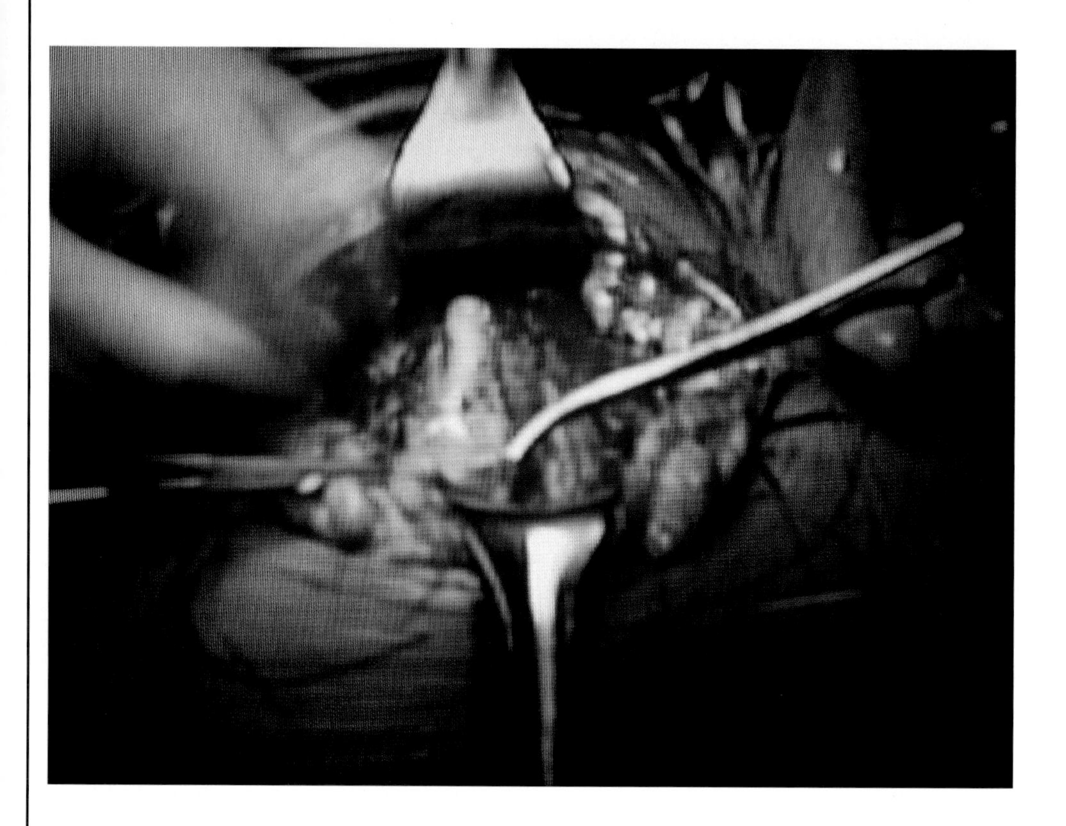

LUTZ BACHER HUGE UTERUS

Lutz writes: This is the real time video record of the operation on my uterus. During the operation the surgeon writes explanatory notes and one of these notes serves as the extended title of the piece: "Huge Uterus with many tumors. No cancer. The tissue is healthy except for tumors. Remove tumors. The uterus is an organ that heals well naturally." Sound is supplied by a relaxation/preparation for your operation tape: "Let yourself now for a moment imagine that you can step out of your body and begin to visualize the procedure as it would look from outside your body. See each step as though you were in the operating room looking over the doctor's shoulder, almost as if you are watching a technician repair an automobile or television, objectively watching and seeing everything going perfectly..." These tapes play on equipment that is configured as body/monitor/hookup.

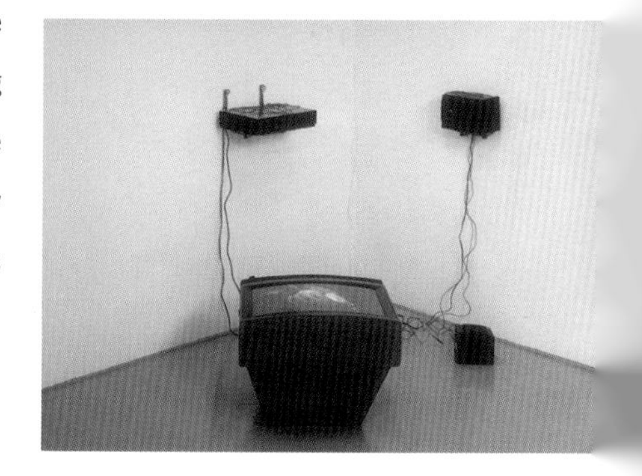

Above: *Huge Uterus,* 1989, detail, color VHS with auxiliary sound, 6 hours.
Right: *Huge Uterus,* installation view, courtesy Simon Watson.

With-jecT
Fabian Marcaccio

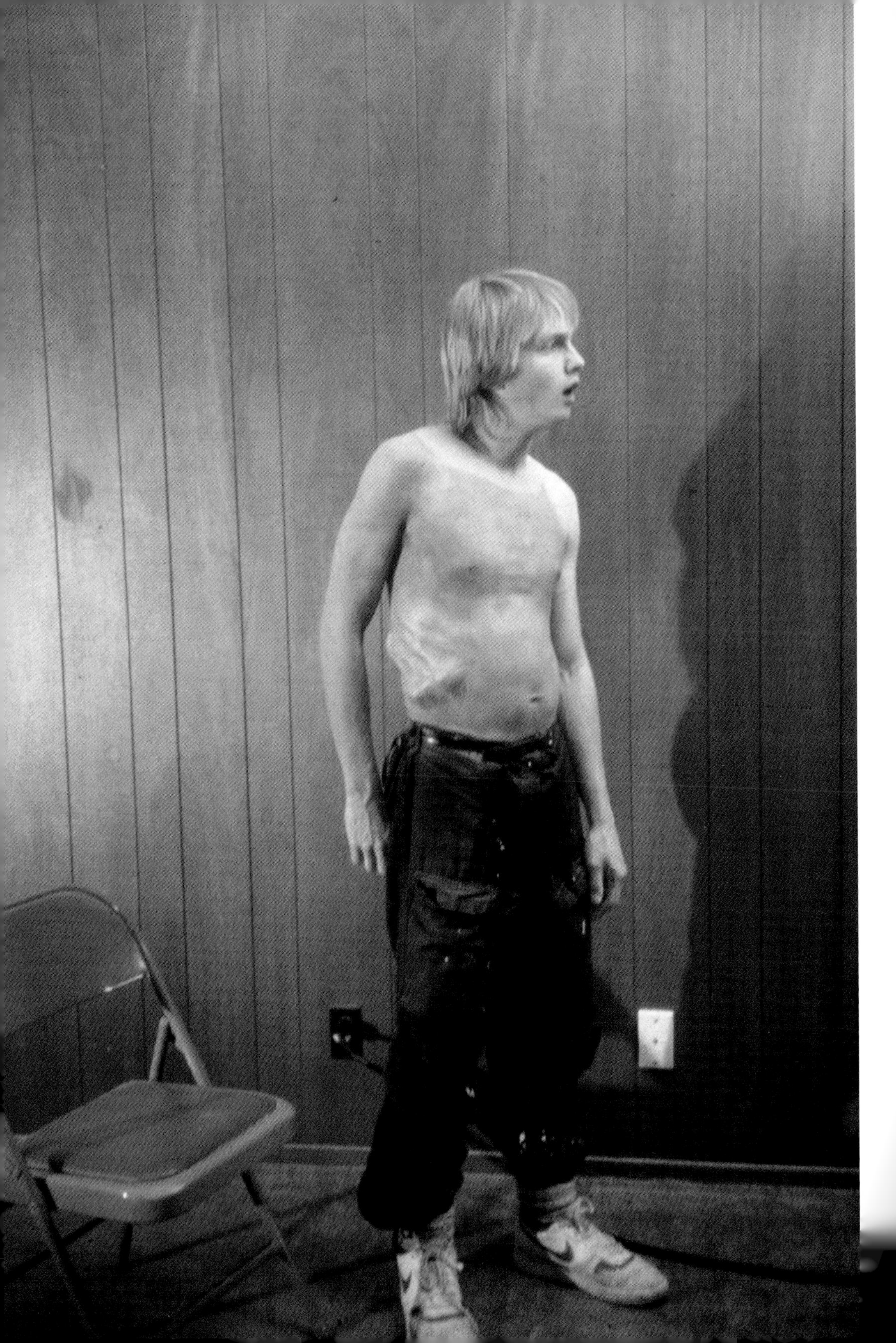

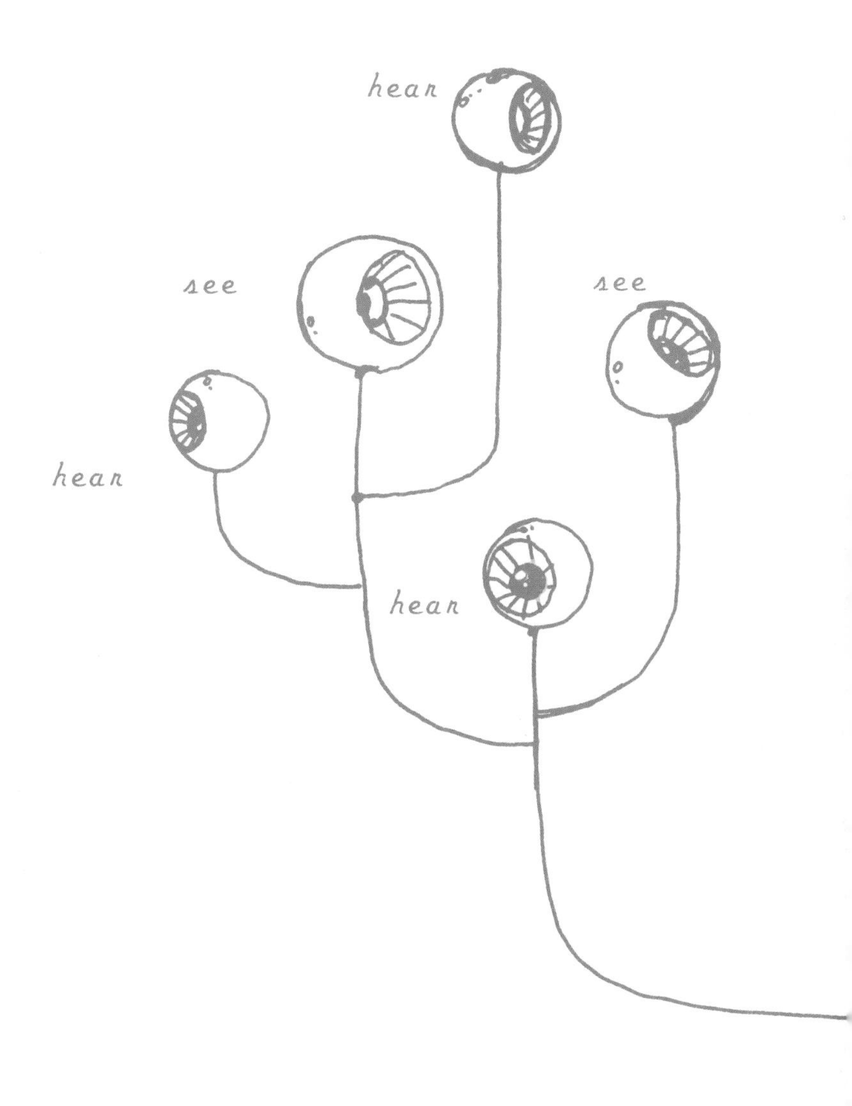

"nice lamp tree..."

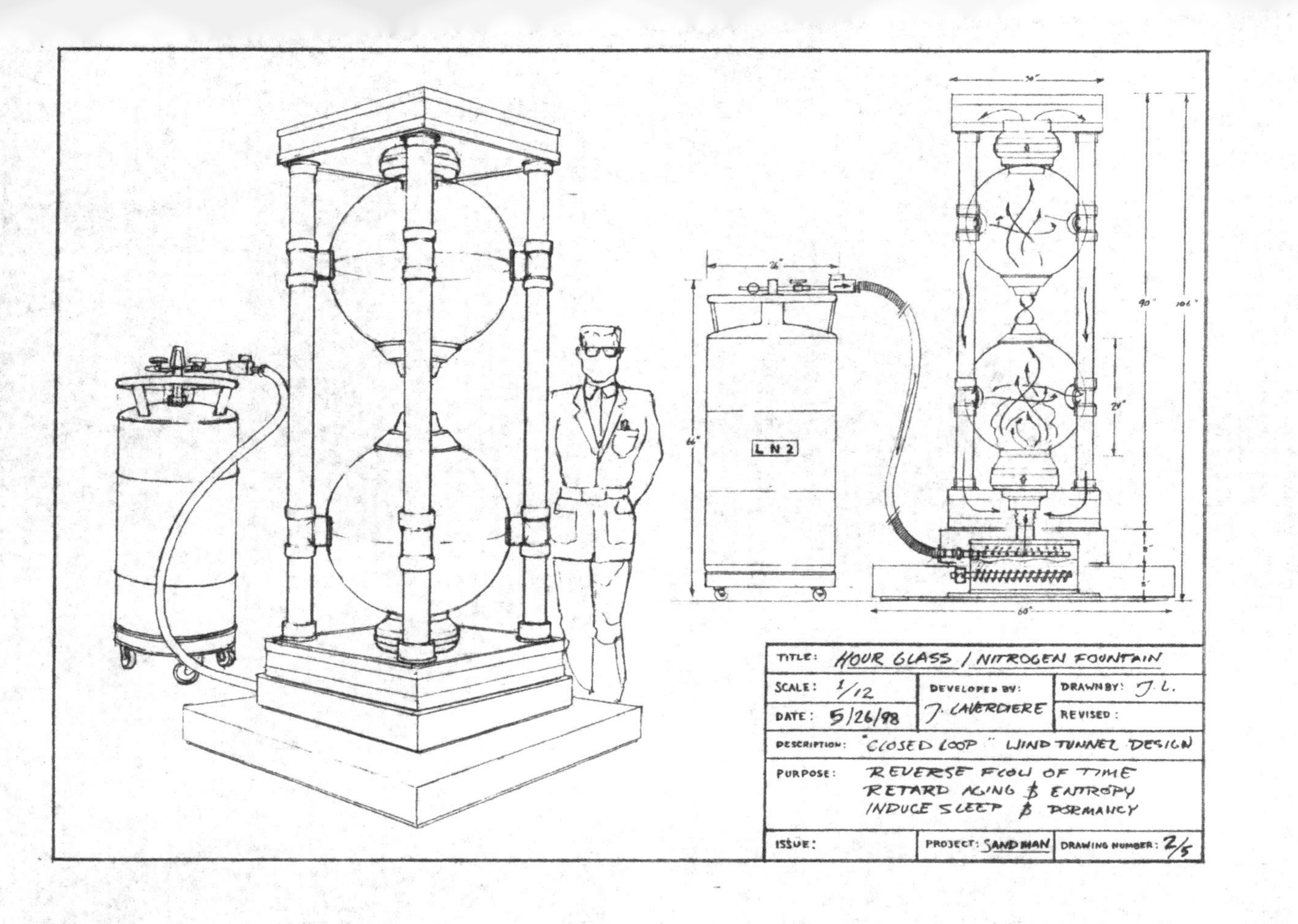

LN2
26"
66"
90"
106"
24"
8"
60"
TITLE: HOUR GLASS / NITROGEN FOUNTAIN
SCALE: 1/12
DATE: 5/26/98
DEVELOPED BY: J. CAVERDIERE
DRAWN BY: J. L.
REVISED:
DESCRIPTION: "CLOSED LOOP" WIND TUNNEL DESIGN
PURPOSE: REVERSE FLOW OF TIME
RETARD AGING & ENTROPY
INDUCE SLEEP & DORMANCY
ISSUE:
PROJECT: SANDMAN
DRAWING NUMBER: 2/5

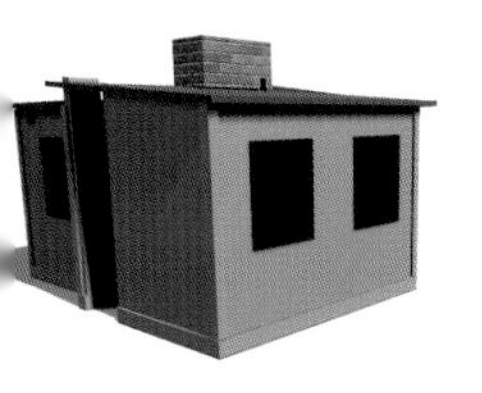

TEAC
TEAC

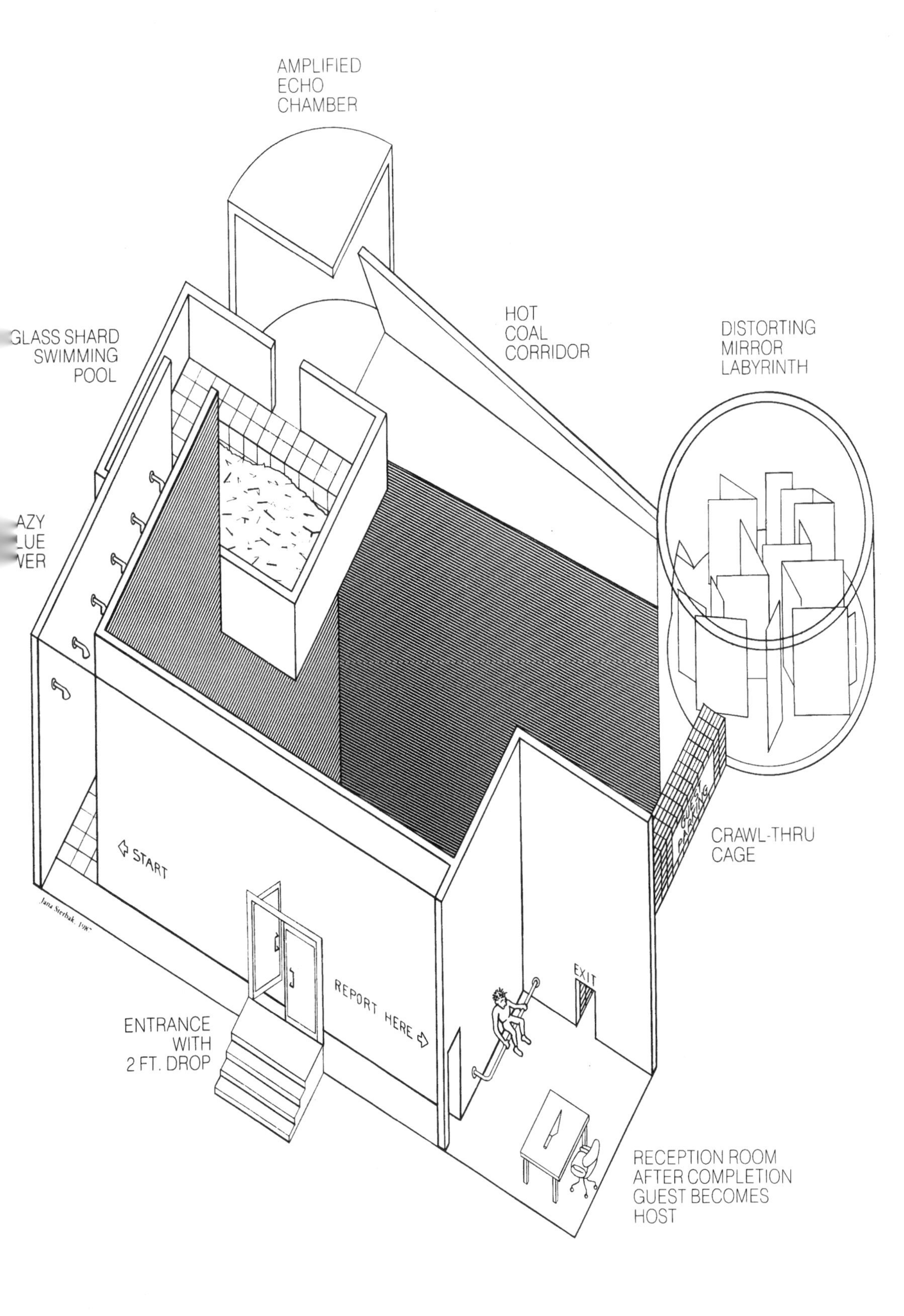
AMPLIFIED ECHO CHAMBER
HOT COAL CORRIDOR
DISTORTING MIRROR LABYRINTH
GLASS SHARD SWIMMING POOL
START
CRAWL-THRU CAGE
GUEST PARKING
Jana Sterbak, 1987
ENTRANCE WITH 2 FT. DROP
REPORT HERE
EXIT
RECEPTION ROOM AFTER COMPLETION GUEST BECOMES HOST

Inside cover
DAVID CRONENBERG still from ***Crash*** **1996**

1 LUISA LAMBRI
Plan Libre (Out of Site) **1997**
cibachrome print; courtesy of the artist

2 LYGIA CLARK
Baba Antropofagico **1973**
documentation of performance; courtesy Museu de Arte Moderna do Rio de Janeiro

3 LAURA PARNES
No Is Yes (Erica, Jen Daking and Joey Friolli) **1998**
photo: Laure Leber; courtesy of the artist

4 MIRANDA LICHTENSTEIN
Untitled **1997**
c-print; courtesy Feature, Inc. and Steffany Martz, New York

5 JASON FOX
Untitled **1997**
marker on paper; courtesy Feature, Inc., New York

6 LUTZ BACHER
Huge Uterus **1989**
color VHS with auxiliary sound, 6 hours, detail and installation view; courtesy Simon Watson

7 CHARLES LONG
Process Photograph **1993**
courtesy of the artist

8 FABIAN MARCACCIO
Environmental Paintant (Remix) **1997-98**
oil and water based paint, silicon gel on canvas and plastic mesh, copper tubing, nylon ropes; courtesy of the artist and Bravin Post Lee, New York

9 JANE & LOUISE WILSON
Crawl Space **1995**
16mm film, installation view, Milch Gallery, London; courtesy Lisson Gallery, London

10 GREG LYNN
Korean Presbyterian Church of New York **1998**
oblique view of the interior sanctuary, computer rendering; courtesy of the artist

11 BRYAN CROCKETT
Nerocrophilia Series **1997**
latex balloons, epoxy resin, glass, polariod photographs, latex tubing, pressure plug, video still; courtesy of the artist

12 KEITH EDMIER
Make-Up Test for "The Fly" 1986 **1998**
c-print; courtesy Friedrich Petzel, New York

13 GREGORY CREWDSON
Untitled **1997**
c-print; courtesy Luhring Augustine, New York

14 ALEXANDER ROSS
Untitled **1998**
color photograph; courtesy of the artist and Feature, Inc., New York

15 JEREMY BLAKE
page from ***Carpetbaggers*** **1997**
ink on paper; courtesy of the artist and Bronwyn Keenan, New York

16 ALEXIS ROCKMAN
Embryo **1996**
envirotex, stainless steel, plasic, plexiglass, oil paint on wood; courtesy Jay Gorney Modern Art, New York

17 JULIAN LA VERDIERE
Hour Glass / Nitrogen Fountain **1998**
blueprint; courtesy of the artist

18 JOSÉ ANTONIO HERNÁNDEZ DÍAZ
Untitled (Nails) **1997**
sand blasted acrylic; courtesy Sandra Gering, New York

19 MARIKO MORI
Mirage **1997**
digital video stills, formated on glass panel; courtesy Deitch Projects, New York

20 SHELLBURNE THURBER
Motel Room with Orange Bedspread **1989**
ektacolor print; courtesy of the artist and Jack Shainman Gallery, New York

21 RANDALL PEACOCK
Artist's Study **1998**
courtesy of the artist

22 TONY OURSLER
Kill or bee... **1998**
mixed media, video projection; courtesy of the artist and Metro Pictures, New York

23 JOHN BRATTIN
The Season of Sadness **1998**
production still; photo: Byrne Guarnotta

24 BONNIE COLLURA
Photograph wearing parts from Destruction Set **1996**
courtesy of the artist and Basilico Fine Arts, New York

25 JANA STERBAK
House of Pain **1987-88**
architectural plan; courtesy Galerie René Blouin, Montréal

28 **MOVING STATUES, FROZEN BODIES**
by **Slavoj Žižek**
from *The Plague of Fantasies*, Verso, 1997

33 **THE SPECTRALIZATION OF THE FETISH**
by **Slavoj Žižek**
from *The Plague of Fantasies*, Verso, 1997

36 **CRONENBERG'S M. BUTTERFLY: THE SPECTACLE OF TRANSVESTISM**
by **Amresh Sinha**

42 **THE SUSPENSION OF THE MASTER**
by **Slavoj Žižek**
from *The Plague of Fantasies*, Verso, 1997

45 **PRELIMINARY DOSSIER FOR CRASH: A NOVEL BY J.G. BALLARD (1974) AND A FILM BY DAVID CRONENBERG (1996)**
by **David Rimanelli**

50 **CRASH – CRONENBERG STYLE**
by **Kathy Acker**
from *Bodies of Work*, 1997

53 **SUBVERSIVE BODILY ACTS**
by **Judith Butler**
from *Gender Trouble*, 1990

55 **VIDEODROME**
by **Scott Bukatman**
from *Terminal Identity*, 1993

61 **THE XEROX DEGREE OF VIOLENCE**
by **Jean Baudrillard**

64 **BLUE DATA**
by **Lia Gangitano**

Our approach to assembling the materials for the publication accompanying the exhibition *Spectacular Optical* was twofold. Since David Cronenberg serves as a point of reference, we selected essays and excerpts that examine his films in relation to topics (for example, the conflation between monstrosity, the body and technology, suburban anxiety and violence, ill-fated scientific experiments, etc.) that are being explored by artists. Yet we also felt the need to broaden the perspective, to take it beyond a single artist or personality. The work in the exhibition is so far-ranging that we included texts from writers not usually associated with visual production as a way to connect the show to a much broader social context. We were responding to the tendency in our culture to subsume the making of art under the guise of a generalized "entertainment", which is how the dominant culture conceals its ideological operation even under the most benign circumstances.

Hence we decided to include Judith Butler's ruminations on how gender is constructed despite Cronenberg's continual detachment from the sexual politics of his works. In the same vein we've decided to publish a series of short excerpts from Slavoj Žižek concerning the fetishization of art production as well as the liberating effects of the demystification of the "fallacy of the phallus" so that the "authority" of the text can be creatively undermined.

We have also included a first-time translation of an essay written by Jean Baudrillard concerning the metamorphosis of violence into "hatred" which he likens to an auto-immune pathology. Throughout the book we wanted to call attention to the collusion between the art object or film and its spectacularity. That is, we wanted to explore how entertainment bedazzles us with an opticality that obfuscates the actual act of looking – as in looking through the spectacle in order to reveal the political ramifications of sight.

Sandra Antelo-Suarez
Michael Mark Madore
May 1998

MOVING STATUES, FROZEN BODIES

by **Slavoj Žižek**

How does psychoanalysis relate to the shift from the traditional authority of wisdom passed on from generation to generation to the reign of expert knowledge – that is, to the predominance of the modern reflected attitude which lacks support in tradition? Psychoanalysis is neither a new version of the return to tradition against the excess of modern reflectivity ("we should open ourselves up to the spontaneity of our true Self, to its archaic, primordial forces" – it was Jung who achieved this anti-modernist inversion of psychoanalysis) nor just another version of the expert knowledge enabling us to understand, and thus rationally dominate, even our most profound unconscious processes. Psychoanalysis is, rather, a kind of modernist meta-theory of the impasse of modernity: why, in spite of his "liberation" from the constraints of traditional authority, is the subject not "free"? Why does the retreat of traditional "repressive" Prohibitions not only fail to relieve us of guilt, but even reinforce it? Furthermore, today the opposition between tradition and expert knowledge is more and more reflectively "mediated": the very "return to traditional Wisdom" is increasingly handled by a multitude of experts (on transcendental meditation, on the discovery of our true Self...).

The exact opposite of this is so-called magic realism in literature, which also presupposes as its background the opposition between the traditional enchanted universe and modernity: magic realism presents the very process of modernization (the arrival of machines, the disintegration of old social structures) from the standpoint of the traditional "enchanted" closed universe – from this viewpoint, of course, modernization itself looks like the ultimate magic...[1] And do we not find something similar in the New Age cyberspace cult which attempts to ground the return to old pagan wisdom in the highest technology? (Perhaps aesthetic postmodernism as such is a desperate attempt to infuse pre-modern enchantment into the process of modernization.) Thus we have a double movement of reflective mediation: (the return to) tradition itself becomes the object of modern expertise; modernization itself becomes the ultimate in (traditional) magic – is this not analogous to the opposition between movement and image, where the movement of life itself is conceived as the magic coming-alive of "dead" images while, simultaneously, the "dead" statue or photo is conceived as the "frozen", immobilized movement of life?

This dialectic of mortification is crucial for our understanding of the underlying phantasmic background of ideological formations. It is deeply significant that photography, the medium of immobilization, was first perceived as involving the *mortification* of the living body. Similarly, the X-ray was perceived as that which renders the "interior" of the body (the skeleton) directly visible. Remember how the media presented Roentgen's discovery of X-rays towards the end of the last century: the idea was that X-rays allow us to see a person who is still alive *as if he were already dead*, reduced to a mere skeleton (with, of course, the underlying theological notion of *vanitas*: through the Roentgen apparatus, we see "what we truly are", in the eyes of eternity...). What we are dealing with here is the negative link between visibility and movement: in terms of its original phenomenological status, movement equals blindness; it blurs the contours of what we perceive: in order for us to perceive the object clearly, it must be frozen, immobilized – immobility makes a thing visible. This negative link accounts for the fact that the "invisible man" from Whale's film of the same name becomes visible again at the very moment of his death: "the person who has stopped being alive exists more fully than when actually alive, moving around before us".[2] Plato's ontology and the Lacanian notion of the mirror-image which freezes motion like a jammed cinema reel overlap here: it is only immobility that provides a firm visible existence.[3]

Against this background, one can establish the contrast between the Gothic motif of a moving statue (or image) and its counterpoint, the inverse procedure of *tableaux vivants*. In his *Elective Affinities*, Goethe provides a nice description of the practice of *tableaux vivants* in eighteenth-century aristocratic circles: famous scenes from history or literature were staged for home amusement, with the living characters on stage remaining motionless – that is, resisting the temptation to move.[4] This practice of *tableaux vivants* is to be inserted into the long ideological tradition of conceiving a statue as a frozen, immobilized living body, a body whose movements are paralysed (usually by a kind of evil spell): the statue's immobility thus involves infinite pain – the *objet petit a* engendered by the stiffness of the living body, its freezing into the form of a statue, is usually a sign of pain miraculously filtered by the statue, from the trickle of blood on the garden statue in Gothic novels to the tears miraculously shed by every self-respecting statue of the Virgin Mary in Catholic countries. The last in this series is the figure of the street entertainer dressed up as a statue (usually as a knight in armour) who remains immobilized for long periods of time: he moves (makes a bow) only when some passer-by throws money into his cup.[5] In contrast to this notion of the statue as a frozen, immobilized body, cinema was perceived at the beginning as the "moving image", a dead image which miraculously comes alive – there in lies its spectral quality.[6] What lurks in the background is the dialectical paradox of the phenomenology of our perception: the immobility of a statue is implicitly conceived as the state of a living being frozen into immobility in an infinite pain; while the moving image is a dead, immobile object which magically comes alive – in both cases, the barrier which separates the living from the dead is transgressed. Cinema is a "moving image", the continuum of dead images which give the impression of life by running at the proper speed; the dead image is a "still", a "freeze-frame" – that is, a stiffened movement.

What we are dealing with here are the two opposed cases of the properly Hegelian paradox of a genus which is its own species – which comprises two species, itself and the species as such. It is incorrect to state that there are two kinds – species of pictures, moving and immobile: the picture "as such" is immobile, frozen, and the "moving picture" is its subspecies, the magic paradox of a "dead" picture coming alive as a spectral apparition. On the other hand, the body as such is alive, moving, and the statue is the paradox of a living body painfully frozen into immobility....[7] A further Lacanian comment to be made is that the primordial point of fixation (or freeze) in what we see is *the gaze itself*: the gaze not only mortifies its object, it stands itself for the frozen point of immobility in the field of the visible. Does not Medusa's head exemplify a gaze which was transfixed when it came too close to the Thing and "saw too much"? In a series of Hitchcock films, the effect of momentary immobilization is produced by the actor's direct gaze into the camera (Scottie in the nightmare sequence in *Vertigo*; the detective Arbogast while he is being slaughtered in *Psycho*; and the unfortunate Fane during his suicidal trapeze act in *Murder*).

Thus horror cuts both ways: what provokes horror is not only the discovery that what we took for a living human being is a dead mechanical doll (Hoffmann's Olympia) but also – perhaps even more – the traumatic discovery that what we took for a dead entity (a house, the wall of a cave...) is actually alive – all of a sudden, it starts to trickle, tremble, move, speak, act with (an evil) intent... So we have, on the one side, the "machine in the ghost" (a ship which sails by itself, with no crew; an animal or a human being which is revealed to be a complex mechanism of joints and wheels), and, on the other, the "ghost in the machine" (some sign of *plus-de-jouir* in the machine giving rise to the effect of "It's alive!"). The point is that both these excesses are desubjectivized: the "blind" machine, as well as the "acephalous" formless life-substance, are two sides of drive (unified in the alien-monster, a combination of machine and slimy life-substance). In literary fiction, one often encounters a person who appears to be just another person within the diegetic space, but is effectively a "No-Man", the desubjectivized horror of the pure drive disguised as a normal individual. Numerous commentators, from Kierkegaard onwards, have pointed out that Mozart's Don Giovanni is actually "characterless", a pure machine-like drive to conquer lacking any "depth" of personality: the ultimate horror of this person resides in the fact that he is not a proper person at all.

Fig. 1 **Fig. 2** **Fig. 3**

This paradox of moving statues, of dead objects coming alive and/or of petrified living objects, is possible only within the space of the death drive which, according to Lacan, is the space between the two deaths, symbolic and real. For a human being to be "dead while alive" is to be colonized by the "dead" symbolic order; to be "alive while dead" is to give body to the remainder of Life-Substance which has escaped the symbolic colonization ("lamella"). What we are dealing with here is thus the split between A and J, between the "dead" symbolic order which mortifies the body and the non-symbolic Life-Substance of *jouissance*.

These two notions in Freud and Lacan are not what they are in our everyday or standard scientific discourse: in psychoanalysis, they both designate a properly monstrous dimension. Life is the horrible palpitation of the "lamella", of the non-subjective ("acephalous") "undead" drive which persists beyond ordinary death; death is the symbolic order itself, the structure which, as a parasite, colonizes the living entity. What defines the death drive in Lacan is this double gap: not the simple opposition between life and death, but the split of life itself into "normal" life and horrifying "undead" life, and the split of the dead into "ordinary" dead and the "undead" machine. The basic opposition between Life and Death is thus supplemented by the parasitical symbolic machine (language as a dead entity which "behaves as if it possesses a life of its own") and its counterpoint, the "living dead" (the monstrous Life-Substance which persists in the Real outside the Symbolic) – this split which runs within the domains of Life and Death constitutes the space of the death drive.[8] These paradoxes are grounded in the fact that, as Freud emphasizes repeatedly, there is *no notion or representation of death in the unconscious*: the Freudian *Todestrieb* has absolutely nothing to do with the Heideggerian *Sein-zum-Tode*. Drive is immortal, eternal, 'undead': the annihilation towards which the death drive tends is not death as the unsurpassable limit of man qua finite being. Unconsciously, we all believe we are immortal – there is no death-anxiety [*Todesangst*] in our unconscious, which is why the very phenomenon of "consciousness" is grounded in our awareness of our mortality.

Kierkegaard's notion of "sickness unto death" also relies on the difference between the two deaths. That is to say, the "sickness unto death" proper, its despair, is to be opposed to the standard despair of the individual who is split between the certainty that death is the end, that there is no Beyond of eternal life, and his unquenchable desire to believe that death is not the last thing – that there is another life, with its promise of redemption and eternal bliss. The "sickness unto death", rather, involves the opposite paradox of the subject who knows that death is not the end, that he has an immortal soul, but cannot face the exorbitant demands of this fact (the necessity to abandon vain aesthetic pleasures and work for his salvation), and therefore, desperately wants to believe that death *is* the end – that there is no divine unconditional demand exerting its pressure upon him:

> *It is not that he cannot, through an unquenchable desire*
> *to overcome the limitation of death, bring his desire*
> *into line with the rational belief that he will not overcome it;*
> *it is rather that he cannot bring his desire*
> *into line with what he fundamentally knows because of*
> *an unquenchable desire to avoid the unpleasantness implicit*
> *in the ability to overcome the limitation.*[9]

The standard religious *je sais bien, mais quand même* is inverted here: it is not that "I know very well that I am a mere mortal living being, but none the less I desperately want to believe that there is redemption in eternal life"; rather, it is that "I know very well that I have an eternal soul responsible to God's unconditional commandments, but I desperately want to believe that there is nothing beyond death; I want to be relieved of the unbearable pressure of the divine injunction." In other words, in contrast to the individual caught in the standard sceptical despair – the individual who knows he will die, but cannot accept it and hopes for eternal life – we have here, in the case of the "sickness unto death", the individual who desperately wants to die, to disappear forever, but knows that he cannot do it: that he is condemned to eternal life. The predicament of the individual "sick unto death" is the same as that of the Wagnerian heroes, from the Flying Dutchman to Amfortas in *Parsifal*, who desperately strive for death, for the final annihilation and self-obliteration which would relieve them of the Hell of their "undead" existence.

1. See Franco Moretti, *Modern Epic* (London: Verso 1998). And was not Twin Peaks a desperate attempt, interesting in its very ultimate failure, to produce a "magic realism" within the very developed First World society? (This idea was suggested to me by Susan Willis.)

2. Paul Virilio, *The Art of the Motor* (Minneapolis: University of Minnesota Press 1995): 69.

3. In contrast to humans, some animals perceive only objects which are moving, and are thus unable to see us if we are absolutely frozen – what we have here is the opposition between pre-symbolic real life, which sees only movement, and the symbolized gaze, which can see only "mortified", petrified objects.

4. In *Selective Affinities*, this immobility of *tableaux vivants* can be read as the metaphor for the very stiffness of the novel's heroes, who are unable to abandon themselves to passion.

5. An interesting accident concerning moving statues occurred in the last years of Socialism in Slovenia: at a crossroads near a small town north of the capital, Ljubljana, a small statue-head of Mary in a shrine allegedly started to move, and even shed tears. The local Communist administrators were delighted; they wanted to exploit the phenomenon economically – to attract religious tourists (perhaps at a sideshow to the flourishing business prompted by the regular appearances of Mary in Medjugorje, in nearby Croatia), construct hotels and leisure centres, and so on. Surprisingly, however, the local Catholic priest was violently opposed to the phenomenon, claiming that it was merely a case of mass hallucination, not a genuine miracle. For that reason, the official Party newspaper attacked him for his "antisocial, noncooperative attitude" – why did he refuse to proclaim it as a miracle, and thereby help the community?

6. Thomas Elsaesser (on whom I draw here) made this point apropos of his well-argued rejection of the standard opposition between Lumière (the first "realist," precursor of *cinéma vérité*) and Méliès (the first "fictionalist", founder of the narrative cinema). A careful observation reveals the tight and detailed narrative structure of Lumière's famous short "documentaries" (most of them involving a clear structure of closure where the end echoes the beginning, etc.). Why was this narrative or fictional aspect of Lumière "repressed"? Because it points towards a possible history of cinema which diverges from the one which was actualized and which we all know: Lumière outlined other possibilities of future development, possibilities which – from today's retroactive teleological reading, which is able to discern in the past only the "germs" of the present – are simply invisible. One of the great tasks of the materialist history of cinema is therefore to follow Walter Benjamin and to read the "actual" history of cinema against the background of its "inherent negation", of the possible alternative histories which were "repressed" and, from time to time, break in as "returns of the repressed" (from Flaherty to Godard...), as in the well-known science-fiction motif of parallel universes where a traveller momentarily goes astray and wanders into another universe.

7. As is well known, at the beginning of cinema, the camera related to the stage precisely as if it were a theatrical stage: it stayed on this side of the barrier that separates the podium from the spectators, registering the action as if it were from the point of view of the theatrical spectator. It took some years for cinema as a specific art form to be born: this occurred at the moment in which the camera transgressed the barrier that separates it from the stage, invaded the actors' space, and started to move among them. One of the possible definitions of cinema is thus that it is a subgenre of the theatre – a theatrical performance in which the spectator, by means of his stand-in (camera), moves forward into the space he observes. In a kind of countermove, modern theatre sometimes endeavours to trespass the frontier between podium and public by surrounding the public with the actors, letting the actors mix with the public.

8. Within the domain of psychoanalysis, the compulsive neurotic provides an exemplary case of the reversal of the relationship between life and death: what he experiences as the threat of death, what he escapes from into his fixed compulsive rituals, is ultimately *life itself*, since the only endurable life for him is that of a "living dead", the life of disavowed, mortified desire.

9. Alastair Hannay, *Kierkegaard* (London: Routledge 1991): 33.

Fig. 1
Vertigo, Directed by A.J. Hitchcock, 1958, Courtesy of MoMA

Fig. 2
Invisible Man, Directed by Jeff Well, 1993, Courtesy of MoMA

Fig. 3
Vertigo, Directed by A.J. Hitchcock, 1958, Courtesy of MoMA

THE SPECTRALIZATION OF THE FETISH
by **Slavoj Žižek**

There is a strong temptation today to renounce the notion of fetishism, claiming that its basic mechanism (the obfuscation of the process of production in its result) is no longer operative in our era of a new kind of "false transparency". The paradigmatic case of this is the recent series of "The making of..." films which accompany big-budget productions: *Terminator 2, Indiana Jones,* and so on: far from destroying the "fetishist" illusion, the insight into the production mechanism in fact even strengthens it, in so far as it renders palpable the gap between the bodily causes and their surface-effect.... In short, the paradox of "the making of..." is the same as that of a magician who discloses the trick without dissolving the mystery of the magical effect. The same goes, more and more, for political campaign advertisements and publicity in general: where the stress was initially on the product (or candidate) itself, it then moved to the effect-image, and now shifts more and more to the making of the image (the strategy of making an advertisement is itself advertised, etc.). The paradox is that – in a kind of reversal of the cliché according to which Western ideology dissimulates the production process at the expense of the final product – the production process, far from being the secret locus of the prohibited, of what cannot be shown, of what is concealed by the fetish, serves as the fetish which fascinates with its presence.

At a somewhat different level, another sign of the same tendency is the fact that today failures themselves have lost their Freudian subversive potential, and are becoming more and more the topic of mainstream entertainment: one of the most popular shows on American TV is "The best bloopers of...", bringing together fragments of TV series, movies, news, and so on, which were cut because something stupid occurred (the actor muddled his lines, slipped...). From time to time, one even gets the impression that the slips themselves are carefully planned so that they can be used in just such a show. The best indicator of this devaluation of the slip is the use of the term "Freudian slip" ("Oh, I just made a Freudian slip!"), which totally suspends its subversive sting.

The central paradox (and perhaps the most succinct definition) of postmodernity is that the very process of production, the laying-bare of its mechanism, functions as the fetish which conceals the crucial dimension of the form, that is, of the social *mode* of production.[1] In a step further in this discussion of Marx, one is thus tempted to propose a schema of three successive figures of fetishism, which form a kind of Hegelian "negation of negation": first, traditional interpersonal fetishism (Master's charisma); then standard commodity fetishism ("relations between things instead of relations between people", that is, the displacement of the fetish on to an object); finally, in our postmodern age, what we witness as the gradual dissipation of the very materiality of the fetish. With the prospect of electronic money, money loses its material presence and turns into a purely virtual entity (accessible by means of a bank card or even an immaterial computer code); this dematerialization, however, only strengthens its hold: money (the intricate network of financial transactions) thus turns into an invisible, and for that very reason all-powerful, spectral frame which dominates our lives. One can now see in what precise sense production itself can serve as a fetish: the postmodern transparency of the process of production is false in so far as it obfuscates the immaterial virtual order which effectively runs the show.... This shift towards electronic money also affects the opposition between capital and money. Capital functions as the sublime irrepresentable Thing, present only in its effects, in contrast to a commodity, a particular material object which miraculously "comes to life", starts to move as if endowed with an invisible spirit. In one case, we have the excess of materiality (social relations appearing as the property of a pseudo-concrete material object); in the other, the excess of invisible spectrality (social relations dominated by the invisible spectre of Capital). Today, with the advent of electronic money, the two dimensions seem to collapse: money itself increasingly acquires the features of an invisible spectral Thing discernible only through its effects.

Again, the paradox is that with this spectralization of the fetish, with the progressive disintegration of its positive materiality, its presence becomes even more oppressive and all-pervasive, as if there is no way the subject can escape its hold...why? Crucial for the fetish-object is that it emerges at the intersection of the two lacks: the subject's own lack as well as the lack of his big Other. Therein lies Lacan's fundamental paradox: within the symbolic order (the order of differential relations based on a radical lack), the positivity of an object occurs not when the lack is filled but, on the contrary, when *two lacks overlap*. The fetish functions simultaneously as the representative of the Other's inaccessible depth *and* as its exact opposite, as the stand-in for that which the Other itself lacks ("mother's phallus"). At its most fundamental, the fetish is a screen concealing the liminal experience of the Other's impotence – the experience best epitomized by the vertiginous awareness that "the secrets of the Egyptians were also secrets for the Egyptians themselves", or (as in Kafka's novels) that the all-pervasive spectacle of the Law is a mere semblance staged in order to fascinate the subject's gaze.[2]

Within the domain of psychoanalytic treatment, this ambiguity of the object which involves the reference to the two lacks becomes visible in the guise of the opposition between the fetish and the phobic object: in both cases we are fascinated, our attention is transfixed, by an object which functions as the stand-in for castration; the difference is that in the case of the fetish, the disavowal of castration succeeds; while in the case of the phobic object, this disavowal fails, and the object directly announces the dimension of castration.[3] Gaze, for example, can function as the fetish-object *par excellence* (nothing fascinates me more than the Other's gaze, which is fascinated in so far as it perceives that which is "in me more than myself", the secret treasure at the kernel of my being), but it can also easily shift into the harbinger of the horror of castration (the gaze of Medusa's head). The phobic object is thus a kind of reflection-into-self of the fetish: in it, the fetish as the substitute for the lacking (maternal) phallus, turns into the harbinger of this very lack.... The point not to be missed is that we are dealing with *one and the same object*: the difference is purely topological.[4] Phobia articulates the fear of castration, while in fetishist perversion (symbolic) castration is that which the subject is after, his object of desire. That is to say: even with the fetishist disavowal of castration, things are more ambiguous than they may seem. Contrary to the doxa, the fetish (or the perverse ritual which stages the fetishist scene) is not primarily an attempt to disavow castration and stick to the (belief in the) maternal phallus; beneath the semblance of this disavowal, it is easy to discern traces of the desperate attempt, on the part of the perverse subject, to *stage* the symbolic castration – to achieve separation from the mother, and thus obtain some space in which one can breathe freely. For that reason, when the fetishist staging of castration disintegrates, the Other is no longer experienced by the subject as castrated; its domination over the subject is complete....

The theoretical lesson of this is that one should invert the commonplace according to which fetishism involves the fixation on some particular content, so that the dissolution of the fetish enables the subject to accomplish the step towards the domain of symbolic universality, within which he is free to move from one object to another, sustaining towards each of them a mediated dialectical relationship. In contrast to this cliché, one should fully accept the paradoxical fact that the dimension of universality is always sustained by the fixation on some particular point.

1. What about Derrida's key criticism of Marx, according to which, in his very probing description of the logic of spectrality in the commodity universe and in social life in general, Marx none the less counts on the revolutionary moment in which the dimension of spectrality as such will be suspended, since social life will achieve complete transparency? The Lacanian answer is that spectrality is not the ultimate horizon of our experience: there is a dimension beyond (or, rather, beneath) it, the dimension of drive attained when one "traverses the fundamental fantasy" (see Slavoj Zizek, "The Abyss Freedom", in F.W.J. Schelling, *The Ages of the World,* Ann Arbor: University of Michigan Press 1997). Furthermore, at this point one should turn the question on Derrida, who himself gets entangled in a necessary ambiguity apropos of the problem of how *la clôture métaphysique* relates to the domain of Western thought. Derrida endlessly varies the motif of how, with regard to this *clôture,* we are neither wholly within nor wholly without, and so on – however, what about Japan or India or China? Are they an inaccessible Outside, and is deconstruction thus constrained to the West, or is *différence* a kind of "universal" structure not only of language, but of life as such, also discernible in animal life?

2. It is easy to discern this redoubled lack already in the functioning of the religious fetish. According to the standard notion, "primitive" religions confuse the material symbol of the spiritual dimension with the spiritual Thing itself: for a primitive fetishist, the fetishized object (a sacred stone, tree, forest) is "sacred" in itself, in its very material presence, not only as a symbol of another spiritual dimension... Does not the true "fetishist illusion" however, *reside in the very idea that there is a (spiritual) Beyond occluded by the presence of the fetish?* Is it not the ultimate sleight of hand of the fetish to give rise to the illusion *that there is something beyond it,* the invisible domain of Spirits?

3. See Paul-Laurent Assoun, *Le Regard et la voix* (Paris: Anthropos 1995): vol. 2: 15.

4. This allows us also to throw new light on the relationship between the two fetish-objects in Richard Wagner: the ring in the *Ring* cycle and the Grail cup in *Parsifal:* the Grail is stable, immovable, *it remains in its place* and shows itself only from time to time, whereas the ring is out of place and *circulates around;* for this reason, the Grail brings incommensurable joy, whereas the ring brings disaster and doom to whomsoever possesses it.... What one has to do, of course, is to assert the 'speculative identity' between the two: they are one and the same object conceived in a different modality.

CRONENBERG'S M. BUTTERFLY: THE SPECTACLE OF TRANSVESTISM
by **Amresh Sinha**

The task of this essay is to find proper ways of providing critical and analytical readings of David Cronenberg's film, *M. Butterfly*, in relation to the various etymological strands, nuances, and hyperbole contained in the word "spectacle". The reader will observe a palimpsest of semiotic, semantic, and syntactic operations designed to spread the semantic register of the word spectacle into a wider and broader site of linguistic transfiguration. Much less has been written or commented on Cronenberg's linguistic high-jinx, so to speak, in relation to the "spectacular optical" imageries of his films.

My purpose, however, is to examine his works from the etymological associations of the words themselves, which form the family of the word spectacle. The most prevalent use of the term, according to the *Oxford English Dictionary*, is to signify a theatrical representation: "a specially prepared or arranged display of more or less public nature forming an impressive or interesting show or entertainment for those viewing it." The French critic, Roland Barthes, considers the theater semiotically privileged, for it is polyphonous in form – unlike language, which is strictly linear. Barthes' emphasis on the visual nature of the etymological root of the term for spectacle is quite instructive. While the English version more often connotes the thing *being* seen, in French the term faithfully preserves its etymological origin, the Latin *spectare*, "to look at," and often connotes the very *act* of seeing, "*le fait de voir*." Thus, spectacle need not signify passivity but may become an active organizing structure.

Fig.

Jonathan Crary, in *Techniques of the Observer*, makes a similar case. According to him, "most dictionaries make little semantic distinction between the word 'observer' and 'spectator,' and common usage usually renders them effectively synonymous." And more so, continues Crary, "unlike *spectare*, the Latin root for 'spectator,' the root for 'observe' does not literally mean 'to look at.' Spectator also carries specific connotations, especially in the context of nineteenth-century culture...namely, of one who is a passive onlooker at a spectacle, as at an art gallery or theater." Crary finds the range of meanings implied in the word *observare* more conducive to his particular task. It means "to conform one's action, to comply with," as in observing rules, codes, regulations and practices. Unlike the passive spectator locked in the gaze of a spectacle, an observer is the one who sees within a prescribed set of possibilities, one who is embedded in a "system of conventions and limitations."[1]

The Spectacle of Spectacle

In its other meaning the word "spectacle" refers to "a person or thing exhibited to, or set before, the public gaze as an object either (a) of curiosity or contempt, or (b) of marvel and admiration." What sort of spectacle is present in the fate of René Gallimard, the abject French diplomat who is hopelessly in love with a Chinese opera singer, Song Liling in the film, *M. Butterfly*? *M. Butterfly* is a "deconstructivist" play, according to the playwright, David Henry Hwang; a political revision of

Puccini's original play, *Madame Butterfly*, about an oriental woman who sacrifices her life for her Western male lover.[2] Hwang's play is based on the real story of a French diplomat, Bernard Boursicot and his Chinese lover for twenty years, Shi Pei Pu. This is a story of a man obsessed with his love for a Chinese diva, unaware that his lover is a man and not a woman, as well as a spy for the Chinese government. It is only when they are both arrested for espionage that Gallimard comes face to face with the revelation of the true sex of his lover. What follows is what Cronenberg calls the transformation, or becoming the other.

How de-politicized the play becomes in the hands of Cronenberg would entail a much longer article. Suffice to say that he misses several opportunities to understand the complex nature of the sexual politics of the play and its relation to what Edward Said has called "Orientalism". Cronenberg flippantly dismisses the issues of sexual politics and colonialism/imperialism in the play as "a particular brand of feminism and sexism."[3] As far as the issues of colonialism and postcolonialism are concerned, he plainly refuses to "buy" the "basic premise that everybody in Western culture is caught up with this mythology of the East."[4] How can one argue with someone who finds questions of historical and sexual repression less important than the emotional power of the theatrical and the operatic?

Is there truly a difference between looking and seeing? As long as we look in order to exercise the power of vision, we are in the vicinity of seeing. However, as soon as we look at being looked at, the power dynamic changes drastically. We look at the other but only in a manner of being seen. Our own look as a directed gaze towards the other is cancelled by the vortex of seeing that looks back at us. We cannot look back. A fundamental asymmetry. René Gallimard cannot look. Every time he insisted on looking, the cultural gaze of the other blocked his view, and he submitted, in his own effrontery, to a cultural codification of what seeing implies. He, perhaps, could only look if he could tear the cultural veil between him and Song Liling, the hymen of looking and seeing, the space which appears as an "in-between."

What is it that veils or hides itself from René Gallimard's look? René Gallimard "looks" at his Butterfly from the moral obligation of a lover, who by financially taking care of his mistress, also fulfills the responsibility, etymologically speaking, of the transitive verb "look," i.e., "to make sure" or "take care," and also "expect" (as in we *look* to have a good year, etc.). Yet, he fails to look into what, precisely, is the blindness that accompanies the predicament of an obsessive love which produces, at the interface of transitivity and intransitivity, the opposite, the failure of both his look and "the ability to see through the real nature of things;" "to ascertain by the use of one's eye," the other meaning and the meaning of the other.

Two different meanings emerge when we attribute the characteristics of spectacle to glass, as it pertains to transparency or making sight better, as in correcting poor vision. It is also a mirror which plays such a decisive role at the end of the film. Gallimard makes a spectacle of himself as he turns into a ridiculous object of derision by failing to differentiate his lover's sex. He not only turns his own life into a spectacle, but he also, paradoxically, turns into a spectator of his own spectacle by fixing his gaze into the mirror of his own reflection, his image as the other. The doubling of spectacle, as a sight and as a specter, has special meaning in this instance. As an object, the spectacle is also a mirror, a piece of glass. However, it is also a deadly weapon, an accessory with which Gallimard slits his throat. Gallimard's suicide turns the spectacle of horror and blood into a special kind of text, a text of metamorphosis, of an allegorical representation of his life's most spectacular transformation into Madame Butterfly. And the *finale* of this act, the spectacle of this immaculate transformation into the image of Madame Butterfly, is not without its own haunting specter, for it is, once again, reflected in the specularity of the mirror as the last image on the primordial screen.

In the end, Gallimard is a victim of the value (in)vested or "speculated," shall we say, in the regime of vision in our technocratic society, that is, the knowledge associated with sight. The word "sight" here corresponds to its double inflections: 1) as a "spectacle" set before the public gaze as an object either of curiosity or contempt and 2) as an ability to see the field of one's vision. People all over the world were, however, intrigued and vexed by a recurring doubt as to how could he not see (as in knowing) the difference, or how could he not feel (as in touching, which ironically also makes the whole affair so touching, so delicate and tragic), the difference of his lover's sex? Why is the knowledge of sight in the history of visual consumption given precedence over the prehistoric sensibility of touch? The epistemological shift between touch and sight, between eschatology and teleology, between Hegel's "sense certainty" and the light of self-consciousness, as it is embodied in the spectacle, is illuminated by a passage in Guy Debord's *Society of the Spectacle*: "The spectacle, as a tendency to make one see the world by means of various specialized mediations (it can no longer be grasped directly), naturally finds vision to be the privileged sense which the sense of touch was for other epochs: the abstract, the most mystifiable sense corresponds to the generalized abstraction of present-day society."[5]

The Spectacle of Transvestism

As Song Liling's identity is revealed, we are once again amidst one of the semantic categories of spectacle: "A thing seen or capable of being seen; something presented to the view, especially of a striking or unusual character; a sight." Indeed, all of the above references to spectacle's various connotations can be espied in the revelation of Song as a man and not a woman. "The sight or view of something." A man or a woman or both! We see *him-her* as a thing, "a sight," presented to our view, because the spectacle is not only staged for Gallimard's humiliation, but for the spectator, who has vicariously and thus surreptitiously, like a spy, identified with the spectacular heterosexual drama. Naked and yet inexplicit: the body of the transvestite does not constitute the real identity. *M. Butterfly* is quite explicit in representing the naked body of Song, though always seen from behind, at the expense of the genitalia. Making a case for sexual identity as a cultural construction, Hwang and Cronenberg refrain from employing the signified of the genitalia, the crux of biological identity, as the determinative force of his (Song's) identity. The transvestite's identity otherwise can only be revealed by means of a doubling or subversion of the cultural signifier, the dress; yet another instance of the arbitrariness of the sign, in this case through male clothing, the Armani suit and tie.

The critique of transvestism, with its emphasis on clothing as a cultural metaphor that cloaks the identity of a transvestite, would benefit enormously from Jacques Derrida's writing on writing in *Of Grammatology*.[6] Derrida states: "Writing, sensible matter and artificial exteriority: a 'clothing.' It has sometimes been contested that speech clothed thought. Husserl, Saussure, Lavelle have all questioned it. But has it ever been doubted that writing was the clothing of speech?

For Saussure it is even a garment of perversion and debauchery, a dress of corruption and disguise, a festival mask that must be exorcised, that is to say warded off, by the good word: 'Writing veils the appearance of language; it is not the guise for language but a disguise.' Strange 'image.' One already suspects that if writing is 'image' and 'exterior figuration,' this 'representation' is not innocent. The outside bears with the inside a relationship that is, as usual, anything but simple exteriority. The meaning of the outside was always present within the inside, imprisoned outside the outside, and vice versa."[7] To Derrida's list, I would add a few from Saussure, who places writing close to insanity; for example: "the tyranny of the written form;" "the phenomenon is strictly pathological;" "a genuine orthographic monstrosity;" and so on.[8] Derrida does not fail to notice the irony of Saussure, the theoretician of the arbitrariness of the sign, so vehemently desiring to form "a natural order of relationships" between speech and writing in order to restore and to recover the natural.

What are the other terrains of transvestism? Is sexuality the only form of inscription to determine its identity, or are the questions posed by identity simply a way of asserting that the discourse must pass through the conceptual determinism of identity to claim both legitimacy and legibility? Transvestism, perhaps, allows us to imagine the possibility of a "double dissymmetry" which goes beyond the codification of sexuality. Not "a-sexual," mind you, says Derrida, but beyond the "binary difference, that governs...all codes,"[9] beyond the feminine/masculine, beyond the chiasmic division of the homo/heterosexual definition. It is possible, via Derrida, to believe in a voice that "may give birth...to another body."[10] Technical devices, prosthetics, make it possible. Voices on the radio and on the telephone, voices that are recorded, no longer belonging to the identity of the body, but, instead functioning as detached, disembodied voices, as writing, as space. The undecidability comes from the writing, where there is writing the identity is rendered undecidable. The marking of sex becomes undecidable in the presence of writing or voice.

Derrida's deconstructive critique reveals the Western world's logocentric bias that privileges the mode of speech over the technique of writing, since the former is supposed to uphold the authority of self-present (spoken) truth. Writing, on the contrary, has been denounced, beginning with Plato's polemical attack in the *Phaedrus* "as the intrusion of artful technique, a forced entry of a totally original sort, an archetypal violence: eruption of the outside within the inside, breaching into the interiority of the soul, the living-presence of the soul within the true logos, the help that speech lends to itself."[11] Saussure, the linguist *par excellence*, "incensed" at writing, states that one has to contend that "although writing is in itself not part of the internal system of the language, it is impossible to ignore this way in which the language is constantly represented. We must be aware of its utility, its defects and its danger."[12] Derrida notes that the kind of "vehemence" Saussure demonstrates in his condemnation of writing "aims at more than a theoretical error, more than a moral fault: at a sort of *stain* and primarily at sin."[13] (emphasis added).

The stain appears everywhere in Cronenberg's work: on the transparent glass door in *The Fly*; on the body of Max Renn in *Videodrome*, after he has "crossed over" and committed the "sin" of witnessing the primal scene, a transgression that leads him to lose his "soul;" and the stain that sullies Gallimard's social reputation. A stain on the culture of French masculinity that relentlessly prides itself on being a connoisseur of love, now confronted with the colossal blunder of a French man who could not distinguish the sex of his own mistress. The stain is also a mark of externality, of contamination and the presence of the virus, which unites the diverse but related discourses of AIDS, homophobia and logocentrism.

The memory of the body dissipates or erases itself from the discursive body, which it leaves open for relentless transcription and projection of meaning, where the authentic self resides in speech. The actual being is located not in the natural part of human existence, the visible form, but in the invisible form, in the voice. Writing appears in the name of voice, as the presence of an absence, the absence of the true self that is present to itself.

Like the transvestite, the specter has a problematic identity, "because the identity of the ghost is precisely the problem." Both constitute a rigorous ontological dilemma, which is haunted by the question of identity. Like the spectral body which is covered head to foot in the armor of its own "costume," the transvestite is also attired in a spectral/spectacular costume; a costume that appears to be the real artifact of the body whose identity is at once protected and rendered problematic (*problema*, Derrida reminds us, is also a shield), precisely because it "prevents perception from deciding on the identity that it wraps so solidly...."[14] The protective armor of the costume acts almost as a prosthetic device that dresses the body in so far as to "dissimulate, protect and mask its own identity."

Marjorie Garber, in "*The Occidental Tourist: M. Butterfly and the Scandal of Transvestitism,*" takes up the question of the cultural production of the transvestite, which is not necessarily limited to the subjects of sexuality (gay or straight) or erotic predilections (dominant or submissive). The very presence of a transvestite, in Garber's terms, "puts in question the cultural representation of gender."[15] Garber's arguments can be summarized as follows: "the transvestite figure...functions simultaneously as a mark of gender undecidability and as indication of category crisis."[16] The charm of transvestism soon gives in to a more trenchant political criticism of Hwang's play, which is categorized as both anti-feminist and openly homophobic. Because of the limitation of space, I am unable to follow in detail the fascinating analyses of Garber's reading of transvestism within the cultural tropism of trespassing and border crossing. The conundrum of the male/female dichotomy has been a central feature of the multiple occurrences of border crossings in *M. Butterfly*. Gender, here, exists only in representation or performance. In addition, transvestism manages to bring acting, espionage, nation and race within the categorical imperatives of the Symbolic. A transvestite's existence depends on the fact of passing for another; a spy's action pertains to the *passing* of information; an actor passes for a character; and in gender there is nothing but this *passing* or performance.

Transvestism and Translation

Transvestism lies beyond the limits of symbolic association. In its rhetorical affinity, transvestism is more closely aligned to the structure of allegorical representation. It is not a thing, and thus, not confined to the status of a thing as identifiable; nor is it a magician's trick which replaces the body with an illusion. But, of course, a certain deception lies at the very core of transvestism, an ambiguity that aligns it to unreality and spectrality. Moreover, transvestism is also a mode of allegorical representation that transcends the question of identity – its *neither nor* structure is open to infinite interpretation, and the ontological structure that forms the back bone of all hermeneutic discourse is also dissimulated. It never loses its ambiguity or its *sovereignty*. As a form of identity,

Fig.

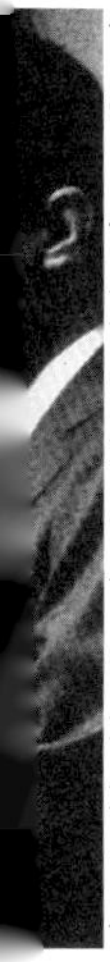

transvestism neither points to this or that. It remains, as Maurice Blanchot might say, (in)different difference, because otherwise all differences are presumed to have fallen in the discourse of identity.

I will conclude by making a few remarks on Walter Benjamin's essay, "The Task of the Translator," in order to illuminate the spectrality of transvestism in relation to translation and language. Continuing with the interweaving trope of "clothing," let us follow Benjamin as he discusses the relationship between the original and the translation: "While content and language form a certain unity in the original, like a fruit and its skin, the language of the translation envelops its content like a *royal robe* with ample folds."[17] (emphasis added). Like translation, transvestism does not relate organically to the body/text that precedes it. Translation of a work of art involves overcoming the natural and symbolic unity rooted in its "mythical interconnectedness" which then becomes "thoroughly denaturalized." Translation approaches the language in order to remove it from its natural associations, to instigate a call for its departure from the "symbolic forest," of which Baudelaire wrote so derisively in his *Spleen* poems. The royal robe of translation/transvestism is a transparent robe, a "see through" garment. It does not "cover" so much as "envelop" the (original) body, "but [it] allows (the light of) the pure language...to shine upon the original all the more fully."[18]

1. Jonathan Crary, *Techniques of the Observer.* (Cambridge: MIT Press, 1996): 5-6.

2. David Henry Hwang, *M. Butterfly, with an Afterword by the Playwright* (New York: New American Library, 1989): 95.

3. Chris Rodley, ed. *Cronenberg on Cronenberg.* (London: Faber and Faber, 1992): 173.

4. Chris Rodley: 173.

5. Guy Debord, *Society of the Spectacle.* (Detroit: Black and Red. 1983).

6. Jacques Derrida, *Of Grammatology.* Trans. Gayatri Chakravorty Spivak. (Baltimore: The Johns Hopkins University Press, 1974): 35.

7. Jacques Derrida, *Of Grammatology*: 35.

8. Ferdinand de Saussure, *Courses in General Linguistics.* Trans. Roy Harris. (La Salle: Open Court, 1972): 31. See also Žižek for whom the stain represents the obscene intrusion of the "Real" that sticks out from the frame of symbolic reality, "a traumatic surplus of the Real over the Symbolic." Slavoj Žižek, ed. *Everything You Always Wanted to Know About Lacan But Were Afraid to Ask Hitchcock.* (London: Verso, 1992): 235.

9. Jaques Derrida, *Points...Interviews,* 1974-1994. Trans. Peggy Kamuf & others. (California: Stanford University Press, 1995): 157.

10. Jacques Derrida, *Points...Interviews*: 161.

11. Jacques Derrida, *Of Grammatology*: 34.

12. Remarking bitterly against "the ascendancy" of writing over speech, Saussure categorically states that the sole criterion for writing to exist is to represent speech, to play a secondary role. Instead, writing has managed to "usurp the principal role" because of its intimate connection to the spoken word. The principal role of language is accorded to speech, for it is speech and not the written word that constitutes a linguist's object of study. Ferdinand de Saussure: 24.

13. Ferdinand de Saussure: 43.

14. Jacques Derrida, *Of Grammatology*: 8.

15. Marjorie Garber, "The Occidental Tourist: *M. Butterfly* and the Scandal of Transvestitism." *Nationalisms & Sexualities.* ed. Andrew Parker, et al. (New York: Routledge, 1992): 123.

16. Marjorie Garber: 125.

17. Walter Benjamin, "The Task of the Translator," in *Illuminations,* Trans. Harry Zohn (New York: Schocken Books, 1969): 75.

18. Walter Benjamin: 79.

Fig. 1, Fig. 2
M. Butterfly, Directed by David Cronenberg, © 1993 Geffen Pictures, photos: Takashi Seida; courtesy of MoMA

Fig. 3
M. Butterfly, Directed by David Cronenberg, © 1993 Geffen Pictures, photo: Takashi Seida; courtesy of Cinematheque Ontario, Toronto

THE SUSPENSION OF THE MASTER
by **Slavoj Žižek**

The supreme example of symbolic virtuality, of course, is that of (the psychoanalytic notion of) castration: the feature which distinguishes symbolic castration from the "real" kind is precisely its virtual character. That is to say: Freud's notion of castration anxiety has any meaning at all only if we suppose that *the threat of castration* (the prospect of castration, the "virtual" castration) *already produces real "castrating" effects*. This actuality of the virtual, which defines symbolic castration as opposed to the "real" kind, has to be connected to the basic paradox of power, which is that symbolic power is by definition virtual, power-in-reserve, the threat of its full use which never actually occurs (when a father loses his temper and explodes, this is by definition a sign of his *impotence*, painful as it may be). The consequence of this conflation of actual with virtual is a kind of transubstantiation: every actual activity appears as a "form of appearance" of another "invisible" power whose status is purely virtual – the "real" penis turns into the form of appearance of (the virtual) phallus, and so on. That is the paradox of castration: whatever I do in reality, with my "real" penis, is just redoubling, following as a shadow, another virtual penis whose existence is purely symbolic – that is, phallus as a signifier. Let us recall the example of a judge who, in "real life", is a weak and corrupt person, but the moment he puts on the insignia of his symbolic mandate, it is the big Other of the symbolic institution which is speaking through him: without the prosthesis of his symbolic title, his "real power" would instantly disintegrate. And Lacan's point apropos of the phallus as signifier is that the same "institutional" logic is at work already in the most intimate domain of male sexuality: just as a judge needs his symbolic crutches, his insignia, in order to exert his authority, a man needs a reference to the absent-virtual Phallus if his penis is to exert its potency.

Swiss bureaucracy provides an illustrative case of this effectivity of virtuality. A foreigner who wants to teach in Switzerland has to appear before a state agency called the *Comité de l'habitant*, and to apply for a *Certificat de bonne vie et moeurs*; the paradox, of course, is that nobody can *get* this certificate – the most a foreigner can get, in the case of a positive decision, is a paper stating that he is *not to be refused* it – a double negation which, however, is not yet a positive decision.[1] This is how Switzerland likes to treat an unfortunate foreign worker: your stay there can never be fully legitimized; the most you can get is the admission which allows you to dwell in a kind of in-between state – you are never positively accepted, you are just not yet rejected and thus retained with a vague promise that, in some indefinite future, you stand a chance....

Furthermore, the very notion of "interface" has its pre-digital precursors: is not the notorious square opening in the side wall of the restroom, in which a gay man (sic) offers part of his body (penis, anus) to the anonymous partner on the other side, yet another version of the function of interface? Is the subject not thereby reduced to the partial object as the primordial phantasmic object? And is not this reduction of the subject to a partial object offered in the interspace-opening also the elementary *sadistic* scene? If, however, the dimension of virtuality and the function of "interface" are consubstantial with the symbolic order, in what, then, *does* the "digital break" consist? Let us begin with an anecdotal observation. As any academic knows, the problem with writing on the computer is that it potentially suspends the difference between "mere drafts" and the "final version": there is no longer a "final version" or a "definitive text", since at every stage the text can be further worked on *ad infinitum* – every version has the status of something "virtual" (con-ditional, provisional).... This uncertainty, of course, opens up the space of the demand for a new Master whose arbitrary gesture would declare some version the "final" one, thereby bringing about the "collapse" of the virtual infinity into definitive reality.

Hackers in California practise a computer manipulation of the *Star Trek* series, so that they add to the "official" TV storyline scenes of explicit sexual encounters without changing any of the "official" content (for example, after the two male heroes enter a room and close the doors, we see a homosexual play between them...). The idea, of course, is not simply to ironize or falsify the TV series, but to bring to light its unspoken implications (the homoerotic tension between the two heroes is clearly discernible to any viewer...).[2] Such changes do not depend directly on technical conditions (the computer's capacity to create lifelike images); they also presuppose the suspension of the function of the Master on account of which – potentially, at least – there no longer is a "definitive version". The moment we accept this break in the functioning of the symbolic order, an entirely new perspective on traditional "written" literature also opens up: why shouldn't we start to produce rewritings of canonic masterpieces to which, without changing the "explicit" content, one would add detailed descriptions concerning sexual activity, underlying power relations, and so on, or simply retell the story from a different perspective, as Tom Stoppard did in his retelling of *Hamlet* from the standpoint of two marginal characters (*Rosencrantz and Guildenstern Are Dead*)? Hamlet itself immediately gives rise to an entire host of ideas: Hamlet is seduced by his mother into incest (or is he himself raping her)? Ophelia kills herself by drowning because she is pregnant by Hamlet? Wouldn't it also be enlightening to rewrite canonic love texts from the feminist standpoint (say, to produce the diaries of the woman who is the object of male advances in "The Diary of a Seducer" from Kierkegaard's *Either/Or*[3])?

In Germany, a whole collection of short stories was written recently, retelling great Western narratives from Oedipus to Faust from the standpoint of the woman involved (Jocasta, Margaretha); even more interesting is the case of the new version of a novel written by a woman and from the romantic woman's perspective, which shifts the focus to another woman – like Jean Rhys's *Wide Sargasso Sea*, which retells Charlotte Brontë's *Jane Eyre* from the standpoint of the "madwoman in the attic", the insane Bertha, Rochester's first wife, imprisoned on the third floor of the Rochester manor house – what we learn, of course, is that, far from simply fitting the category of evil destroyer, she was herself the victim of brutal circumstances....

Since the writer who exemplifies restraint and reliance on the unspoken is definitely Henry James, in whose work tragedies occur and whole lives are ruined during what appears to be a polite dinner-table conversation, would it not also be enlightening to rewrite his masterpieces in order to explicate their latent sexual tensions and political content (Strether from *The Ambassadors* masturbating late in the evening in his hotel room – or, better still, engaged in homosexual play with a paid young boy – to relax after his busy daily social round; Maisie from *What Maisie Knew* observing her mother's lovemaking with her lover)? Once the dam of the

Master-Signifier collapses, it opens up the way for the flood of ideas, some of which can be not only amusing but also insightful in bringing to light the underlying "repressed" content. The problem is, however, that one should also not lose sight of what gets *lost* in such a procedure: it relies on the transgressive move of violating the boundaries of some canonic work – once the canonic point of reference loses its strength, the effect changes completely. Or – to put it in a different way – the effect of some content is entirely different if it is only hinted at as the "repressed" secret of the "public" storyline, as opposed to being openly described.

Franz Kafka's *The Castle* describes the hero's (K.'s) desperate attempts to get into the Castle, the mysterious seat of power. A new CD-ROM, *The Castle*, turns Kafka's novel into an interactive game: the player is invited to guide the hapless K. past Klamm, the mysterious gatekeeper, and into the dark and dank corridors of the castle.... The point here is not to deplore the vulgarity of this idea but, rather, the opposite: to emphasize the structural analogy between K.'s endless attempts to enter into contact with the Castle and the never-ending feature of the interactive computer game, as if that which was, in the case of Kafka, a nightmarish experience turns all of a sudden into a pleasurable game: nobody really wants to enter the Castle fully; the pleasure is provided by the endless game of gradual and partial penetrations. In other words, nightmare turns into pleasurable game the moment the function of the Master is suspended.

The decline of this function of the Master in contemporary Western societies exposes the subject to radical ambiguity in the face of his desire. The media constantly bombard him with requests to choose, addressing him as the subject *supposed to know what he really wants* (which book,clothes, TV programme, holiday destination...): "Press A if you want this; press B if you want that"; or – to quote the slogan of the recent "reflective" TV publicity campaign for advertisement itself – "Advertisement – the right to choose". At a more fundamental level, however, the new media deprive the subject radically of the knowledge of what he wants: they address a thoroughly malleable subject who has constantly to be told what he wants – that is, the very evocation of a choice to be made performatively creates the need for the object of choice. One should bear in mind here that the main function of the Master is to tell the subject what he wants – the need for the Master arises in answer to the subject's confusion, in so far as he does not know what he wants. What happens, then, in the situation of the decline of the Master, when the subject himself is constantly bombarded with the request to give a sign of what he wants? The exact opposite of what one would expect: it is when there is no one there to tell you what you really want, when all the burden of the choice is on you, that the big Other dominates you completely, and the choice effectively disappears – is replaced by its mere semblance. One is again tempted to paraphrase here Lacan's well-known reversal of Dostojevsky ("If there is no God, nothing at all is permitted"): if no forced choice confines the field of free choice, the very freedom of choice disappears.[4]

1. I owe this information to John Higgins, Cape Town University (private conversation).

2. I owe this information to Constance Penley, UCLA (private conversation).

3. Incidentally, Kierkegaard *did* plan also to write "hetaira's diary", a diary of seduction from the perspective of the seductress (who is, typically, conceived of as a "hetaira", i.e. a prostitute).

4. However, one is also tempted to claim that there is a way in which, in cyberspace, the foreclosed dimension of the symbolic Master "returns in the Real": in the guise of supplementary characters who exist only as programmed entities within the cyberspace (like Max Headroom from the TV series of the same name – see Chapter 6 of Stone, *The War of Desire and Technology*). Are not such figures the exemplary cases of what Lacan calls *l'Un-en-plus*, the One which adds itself to the series, the direct point of subjectivization of the anonymous order which regulates relations between "real" subjects?

PRELIMINARY DOSSIER FOR CRASH:
A NOVEL BY J.G. BALLARD (1974) AND A FILM BY DAVID CRONENBERG (1996)
by **David Rimanelli**

1

The technologization of the human body – and its consequences for human sexuality.

2

The ambiguity/ambivalence of sexual identification and sexual preference (as man or woman, for man or woman).

3

The political statement embedded in this confusion: this perversion. Redefinition: In this era of identity politics, words such as "queer" and "nigger" are *reclaimed* by their respective "communities" as positive, valorizing epithets, as well as challenges and curses aimed at Normal Life and its self-proclaimed progressive mouthpieces. Blaxploitation is back: Bad-Nigger-wannabe white-guy director (of genius) Quentin Tarantino's *Jackie Brown*, certainly one of the best films of 1998, heroizes the Über-Bad-Nigger character played by Samuel L. Jackson and reinvigorates 70s blaxploitation queen Pam Greer's stardom, the new "mature" black-lady sexuality that will perhaps now supersede Tina Turner, spokeswoman for Hanes Pantyhose.

Boredom: the casual, yet excruciatingly *ralenti* gesture, such as the smoking of a cigarette or picking up a pair of shears – Cronenberg dwells on these details with the professionally jaded eye of an expert fashion photographer, resuscitating beauty by recourse to the sixties avantgardist aesthetics of real time (how long to pick up those shears, how exquisitely protracted Catherine's drag on a cigarette). And then the rapid reversal of boredom and torpor into speed and violence: the highway, the car crash. Let's get pretentious: The dialectics of desire and repulsion, organic and inorganic, the woman's voluptuous body and the mangled remains of the car crash, acceleration and inertia.

How many times have you characterized a bad sexual experience or a bad relationship as "a car crash"?

Social perversion: How many times have you deliberately drawn a friend or a stranger into a situation of gross discomfort without in any formal way transgressing the codes of politesse? – as Gabrielle does when with mordant yet sweet-natured relentlessness as she traps a gray-suited car salesman into "helping" her into the car she and Ballard are inspecting? The Salesman the hypostasis of uptight middle-classness, Gabrielle that of Differently-Abled-Nymphomaniac Lib. And Ballard that of voyeur (or author).

Speed. You needn't read Paul Virilio to pick up on its metaphors and iconography. Everyone who has ever driven on the freeway knows; everyone who has ever experienced "road rage" knows. (Road rage is nowadays recognized as a major psychiatric disorder as well as a major cause of death.)

8

Back to sex via speed: a quickie is typically understood as a brief encounter (bedroom, boardroom, public toilet, elevator, ditch, car) – often imbued with a negative connotation.

I think now of the other crashes we visualized, absurd deaths of the wounded, maimed, and distraught. I think of the crashes of psychopaths, implausible accidents carried out with venom and self-disgust, vicious multiple collisions contrived in stolen cars on evening freeways amid tired office-workers. I think of the absurd crashes of neurasthenic housewives returning home from their VD clinics, hitting parked cars in suburban high streets. I think of the crashes of excited schizophrenics colliding head-on into stalled laundry vans in one-way streets; of manic-depressives crushed while making pointless U-turns on motorway access roads; of luckless paranoids driving at full speed into the brick walls at the ends of known culs-de-sac; of sadistic charge nurses decapitated in inverted crashes on complex interchanges; of lesbian supermarket manageresses burning to death in the collapsed frames of their midget cars before the stoical eyes of middle-aged firemen; of autistic children crushed in rear-end collisions, their eyes less wounded in death; of buses filled with mental defectives drowning together stoically in roadside industrial canals.
(J.G. Ballard, *Crash,* 1974)

Take note of the two uses in the above passage of words related to the Classical philosophy of stoicism viz. "the stoical eyes of middle-aged firemen"and "drowning together stoically in roadside industrial canals." Think about (read) the Plinys, Elder and Younger; Seneca; Greek antecedents; Christian variations (cf. Boethius, *The Consolation of Philosophy*); Renaissance stoicism (e.g., Montaigne, perhaps especially in *The Apology for Raymond Sebond*, but also in his extensive, indeed relentless quotations from the Stoic philosophers and his attendant commentaries in the *Essays*). Neo-classicism: in Britain, Pope, Addison & Steele, Sir Joshua Reynolds' *Discourses*; the ambiguous transition to proto-Romanticism via Dennis, Burke; the Sublime; the Gothick. In Germany: Lessing, Winckelmann, Goethe. Fill in the literary and art historical etceteras. Further research will be required.

Andy Warhol. The Disaster Paintings. Need we say more. e.g., *Optical Car Crash* (1962); *Suicide (Fallen Body)*, (1963); *Foot and Tire* (1963); *Ambulance Disaster* (1963); *Five Deaths Seventeen Times in Black and White* (1963); *Saturday Disaster* (1964); *Hospital* (1963); *White Burning Car III* (1963).

Relevant quotations of Warhol (and there are many others):

a) If you want to know about Andy Warhol, just look at the surface: of my paintings and films and me, and there I am. There's nothing behind it.
(Gretchen Berg, "Andy: My True Story," *Los Angeles Free Press,* March 17, 1967)

b) I see everything that way, the surface of things, a kind of mental Braille, I just pass my hands over the surface of things.
(ibid.)

c) I like boring things. I like things to be exactly the same over and over again.
(G.R. Swenson, "What Is Pop Art? Answers from 8 Painters, Part I," *Artnews 62,* November 1963)

d) I think everybody should be a machine. I think everybody should like everybody.
(ibid.)

e) I still care about people but it would be so much easier not to care. I don't want to get too close; I don't like to touch things, that's why my work is so distant from myself.
(Read by Nicholas Love, April 1, 1987)

f) If everybody's not a beauty, then nobody is.
(The Philosophy of Andy Warhol *[From A to B and Back Again]*, 1975)

g) I guess it was the big plane crash picture in the front page of a newspaper: *129 DIE*. I was also painting the *Marilyns*, I realized that everything I was doing was Death. It was Christmas or Labor Day – and every time you turned on the radio they said something like "4 million are going to die." That started it. But when you see a gruesome picture over and over again, it doesn't really have any effect.
(Swenson, "What Is Pop Art?")

h) The best atmosphere I can think of is film, because it's three-dimensional physically and two-dimensional emotionally.
(*The Philosophy of Andy Warhol*)

i) I never read, I just look at pictures.
(Andy Warhol, Kaspar König, Pontus Hultén, Olle Granath, eds., *Andy Warhol*, 1968)

j) But I always say, one's company, two's a crowd, three's a party.
(*Andy Warhol's Exposures*, 1979)

k) Truman says he can get anyone he wants. I don't want anyone I can get.
(*ibid.*)

Consider the above quotations with respect to the interpersonal relations of Ballard, Catherine, Vaughan, Dr. Helen Remington, Gabrielle, Colin Seagrave, and also with respect to their interpersonal-objective relations with things: cars, highways, clothes (often fetishistic), prostheses (ditto), cigarettes, shears, jets, etc. Think about some of the above quotations with respect to, say, Catherine's (Deborah Kara Unger's) seamless, white alabaster, opaque, unisubstantial body: A pure surface over which many hands pass.

When I was an undergraduate, I had the privilege of taking a graduate seminar in Wordsworth and Coleridge with a great scholar and teacher, Geoffrey Hartman. Professor Hartman counseled – no, warned – his students to resist the rush to thematization in a famous and apparently straightforward sonnet by Wordsworth. Resist the rush to thematization with respect to Andy Warhol's statements and his art works: resist irony. Try taking it at face value. What you see (or read) is what you see (and read). Sincerity. Honesty is the ultimate act. But can you pull it off?

[...] a link can be made with several Warhol series that use photographs of automobile accidents. These commemorate events in which the supreme symbol of consumer affluence, the American car of the 1950s, lost its aura of pleasure and freedom to become a concrete instrument of sudden and irreparable injury. (In only one picture of the period, *Cars*, does an automobile appear intact.) Does the repetition of *Five Deaths* or *Saturday Disaster* cancel attention to the visible anguish of the living of the horror of the limp bodies of the unconscious and dead? One cannot penetrate beneath the image to touch the true pain and grief, but their reality is sufficiently indicated in the photographs to expose one's limited ability to find an appropriate response. As for the repetition, it might just as well be taken to register the grim predictability, day after day, of more events with

an identical outcome, the levelling sameness with which real, not symbolic, death erupts into daily life.
(Thomas Crow, "Saturday Disasters: Trace and Reference in Early Warhol," [1987; revised 1990])

Vaughan died yesterday in his last car-crash. During our friendship he had rehearsed his death in many crashes, but this was his only true accident. Driven on a collision course towards the limousine of the film actress, his car jumped the rails of the London Airport flyover and plunged through the roof of a bus filled with airline passengers. The crushed bodies of package tourists, like a haemorrhage of the sun, still lay across the vinyl seats when I pushed my way through the police engineers an hour later. Holding the arm of her chauffeur, the film actress Elizabeth Taylor, with whom Vaughan had dreamed of dying for so many months, stood alone under the revolving ambulance lights. As I knelt over Vaughan's body she placed a gloved hand to her throat.
(J.G. Ballard, *Crash*)

In Cronenberg's film version, the iconographic and metaphorical possibilities of certain fashion accoutrements, for example gloves, in particular those of Dr. Helen Remington, but also those of Catherine. Calling Max Klinger.

O love, O celibate.
Nobody but me walks the waist-high wet.
The irreplaceable
Golds bleed and deepen, the mouths of Thermopylae.
(Sylvia Plath, "Letter in November," *Ariel*, copyright 1961-65, by Ted Hughes)

James Ballard and his wife, Catherine, are locked into a practice of compulsive sex with strangers. They compare notes, seeking any physical experience that makes sense in a bleak, passionless world of multi-lane freeways. Ballard becomes involved with Helen Remington after he accidentally ploughs into her car, killing her husband. Their mutual crash-victim status, ultimately delivering them into the sump-oiled world of the pathological Vaughan – a renegade scientist and leader of a strange subterranean group who spend their time looking at videos of simulated car crashes, fucking in cars or attending Vaughan's "illegal" performances, such as his restaging of James Dean's 'Death by Porsche'. Ballard, Catherine, and Helen Remington are all drawn into Vaughan's crazed orbit, and his dream of a new conceptualized relationship of flesh and metal: man and machine. When Vaughan is killed in a motorway car crash, Ballard takes his place and the cycle continues. The film ends with Ballard caressing Catherine after he has forced her car off the motorway.
(Synopsis of *Crash,* in *Cronenberg on Cronenberg*, ed. Chris Rodley, [1992; revised edition 1997])

David Cronenberg: "Some potential distributors said, 'You should make them more normal at the beginning so that we can see where they go wrong.' In other words, it would be like a *Fatal Attraction* thing. Blissful couple, maybe a dog and a rabbit, maybe a kid. And then a car accident introduces them to these horrible people and they go wrong. I said 'That isn't right, because there's something wrong with them right *now*. That's why they're vulnerable to going even further.' The novel is uncompromising in that way. Why shouldn't the movie be?"
(*ibid.*)

21

– I should have gone to the funeral.
I wish I had. They bury their dead so quickly. They should leave them lying around for months.
What about his wife, the woman doctor. Have you been to visit her yet?
No. I couldn't. I feel too close to her.
(Dialogue between Ballard and Catherine in the film version)

– How can you drive, James? You can barely walk.
Is traffic heavier now. There seem to be three times as many cars as there were before the accident.
I have to leave for work.
(ibid.)

The emphasis on anal penetration, even in the context of nominally heterosexual relations (e.g. Catherine imagining Ballard's sex with Vaughan as he [ambiguously] penetrates [makes love to] her from behind). Bruce Nauman: Run From Fear/Fun From Rear.

Catherine pressed her head into the pillow, a familiar gesture of concentration.
'Do you like Vaughan?'
I moved my fingers to her nipple again and began to erect it. Her buttocks moved on to my penis.
Her voice was pitched on a low, thick note.
'In what way?' I asked.
'He fascinates you, doesn't he?'
'There is something about him. About his obsessions.'
'His flashy car, the way he drives, his loneliness. All the women he's fucked there.
It must smell of semen...'
'It does.'
'Do you find him attractive?'
I drew my penis from her vagina and placed the head against her anus, but she pressed it back into her vulva with a quick hand.
'He's very pale, covered with scars.'
'Would you like to fuck him, though? In that car?'
I paused, trying to delay the orgasm rushing like a tidal race up the shaft of my penis.
'No. But there is something about him, particularly as he drives.'
'It's sex – sex and that car. Have you seen his penis?'
[...]
'Is he circumcised?' Catherine asked. 'Can you imagine what his anus is like? Describe it to me.'
(J.G. Ballard, *Crash*)

"Were there any limits to Vaughan's irony?"
(ibid.)

CRASH – CRONENBERG STYLE

by **Kathy Acker**

Last week I had the privilege of seeing *Crash*, David Cronenberg's film adaptation of James Ballard's novel. Unfortunately for me, I've only seen the film once; moreover, it could be that the distributors will pressure Cronenberg to edit out some of the more graphic sexual shots. I'm only guessing. Nevertheless I want to write about this movie now because what I saw was a masterpiece.

The film is not so much a conventional adaptation as a marriage between two of the most important artists working today, the kind of marriage that can take place only in the world depicted in and heralded by the movie itself, a world in which life and death collide and become one.

Whereas Ballard's novel begins in death, with the death of Vaughan, a car-crash fetishist and scientist, Cronenberg's movie begins with shot after shot of metal. Beautifully painted, gleaming metal. Kenneth Anger composed *Scorpio Rising*, a love song to motorcycles; this is Cronenberg's paean to cars. Just as husband and wife mirror each other, at least in alchemical terms, Cronenberg's sexual reiteration of metal images is a reenactment of Ballard's beginning-in-death.

These very first moments of the movie introduce a world in which desire equals death and vice-versa. In which death, and the approach to human death, is no longer an end but a beginning.

In the next paragraph of his novel, Ballard delves further into this new world: "...crushed bodies...like a haemorrhage of the sun." By using this image, he recalls Georges Bataille's sun-spots (sic), splintering of the eye, the pineal eye, the asshole. Ballard turns to a surrealist, even an excommunicated one, the way Cronenberg does to the Anger of *Scorpio Rising*. Both do this in order to make clear that the territory of the car crash and of the crashed body is the realm of ecstasy.

This realm is future insofar as it is other; like Sterling's future, it has a presence/present we haven't yet acknowledged. Absolutely unlike Sterling's, it is a reality shaped by desire, a reality in which desire, especially sexual desire, turns into ecstasy.

Though Cronenberg's film, like Ballard's novel, has a simple narrative, poetry weighs on the narrative until its story is almost nonexistent.

What do I mean by "poetry"? I once asked a playwright, "What is the basis of a play, of theater?" I was trying to figure out how to write a play; I knew I didn't have a clue. He answered: Struggle. There has to be a protagonist and an antagonist because it's the tension between these two that constitutes drama. In *Crash*, there are no protagonists nor antagonists nor psychological drama: there is simply desire. The desire of the characters, the desire of the filmmaker. Every color in the film, every object as it is placed in space and next to other objects, the way that humans are seen as objects, how each frame moves to the next announces and repeats this desire. That is what I mean by "poetry." In Cronenberg's movie, all is crystal clear: the colors, the storyline, above all the intentions and emotions of the characters, a fictional

Ballard, the filmmaker's intentions or desires. The fictional Ballard's longings and perceptions repeat Cronenberg's. It is repetition, not linear narrative, that shapes *Crash*; repetition, as Gertrude Stein noted, is the language of sexual love.

The name of the territory depicted by *Crash* is "violence," but then so is the name of this society in which I'm living. Art is metamorphosis: Cronenberg has transmuted my violent society into a world in which I want to be alive, in which I want to be human.

How has he done this? How can a world that is viable, unlike the nonviable society I know, be created? How can I, can we make our existences viable? In his novel, Ballard speaks of "the celebration of wounds." He states: "For him [Vaughan] these wounds were the keys to a new sexuality born from a perverse technology." According to my lapsing memory, in the movie, the fictional Ballard says something like this to Vaughan, says that he and Vaughan and their partners and lovers, almost all car crash survivors, are recasting human sexuality and so, remodulating the human body in the site or space of the meeting of human and metal, human and technology. Sterling states something similar, but in his novel, in terms of form and content, there is no ecstasy. No, Vaughan replies to the fictional Ballard, that's too simple, that's only a beginning.

How can I, can we make the new world? How can we make our world new? Further in his novel, Ballard comments that death has "jerked loose" the sexual possibilities of everything around him. As Cronenberg began the film with metal shot after metal shot, his camera continues its sexual path by caressing each dismembered car crash limb, each far-flung body. The camera transforms each wound into a new, never-before-seen genitalia. After a car crash, anything can be penetrated, anyone, everything and everyone is, anyone can penetrate and does. This new realm is no longer one of duality, of men who penetrate with cocks and women who get penetrated via cunts. Each car crash allows sexual organs to proliferate everywhere: the world is sexualized as it was when it began and in its constant beginnings.

How can the new world begin? It is the car crash, in other words, it is death that allows us to transmute. I am in this society living in a reality dominated by failed attempts to reject death, to deny death's presence, while more and more of my friends are dying. *Crash*, the film perhaps even more clearly than the book, is the battle cry of those who are remaking sexuality, remaking our perceptions. A battle cry and a slap in the face of rigid liberalism, of those who refuse to acknowledge the body, sex, and death. As I watched *Crash*, I watched my own desires and the ways I desire, and I watched these change.

How can a film remake sex and sexuality if sexual desire lies outside human control? Perhaps the technological society, as in Sterling, believes that the body and sex can be controlled. But then, what of death? This is one reason that sex must always be viewed in partnership with death. I'll give you one example of what Cronenberg does in *Crash*. For me, the central shot of the film, a few seconds long, is one of a penis, I remember that it's Vaughan's, next to Katherine's, Ballard's wife's, cunt. The image immediately repeats itself as a finger in the same position on the cunt. What interested me most was that, contrary and probably antagonistic to all porn conventions, the cock is not hard. Through sexual desire, both his own and that of his characters, Cronenberg has reenvisioned the dominant and always rigid phallus of the old

Fig. 1

the king-must-not-die world as other, soft, another body part, by the end of the film as metal, a car, death, a kiss. In the last moments of the movie, the fictional Ballard promises his wife that he will be able to enact her most profound sexual desires by killing her in a car crash. A slap in the face of the white Western world.

Making, the reenvisioning of sexuality, the making of a new world is personal. Ballard's novel is a love letter to Vaughan. "He knew that as long as he provoked me with his own sex, which he used casually as if he might discard it for ever," says Ballard, "I would never leave him." In Cronenberg's film, Vaughan and the fictional Ballard have sex; the whole film, actually, is Cronenberg's fucking of Ballard. To desire to do is to do and to do is to desire to do in this world which is no longer dualistic. In *Crash*, Cronenberg has made the gods.

Fig. 1
Crash, Directed by David Cronenberg

SUBVERSIVE BODILY ACTS
by **Judith Butler**

If the body is not a "being," but a variable boundary, a surface whose permeability is politically regulated, a signifying practice within a cultural field of gender hierarchy and compulsory heterosexuality, then what language is left for understanding this corporeal enactment, gender, that constitutes its "interior" signification on its surface? Sartre would perhaps have called this act "a style of being," Foucault, "a stylistics of existence." And in my earlier reading of Beauvoir, I suggest that gendered bodies are so many "styles of the flesh." These styles all never fully self-styled, for styles have a history, and those histories condition and limit the possibilities. Consider gender, for instance, as *a corporeal style*, an "act," as it were, which is both intentional and performative, where "*performative*" suggests a dramatic and contingent construction of meaning.

Witting understands gender as the workings of "sex," where "sex" is an obligatory injunction for the body to become a cultural sign, to materialize itself in obedience to a historically delimited possibility, and to do this, not once or twice, but as a sustained and repeated corporeal project. The notion of a "project," however, suggests the originating force of a radical will, and because gender is a project which has cultural survival as its end, the term *strategy* better suggests the situation of duress under which gender performance always and variously occurs. Hence, as a strategy of survival within compulsory systems, gender is a performance with clearly punitive consequences. Discrete genders are part of what "humanizes" individuals within contemporary culture; indeed, we regularly punish those who fail to do their gender right. Because there is neither an "essence" that gender expresses or externalizes nor an objective ideal to which gender aspires, and because gender is not a fact, the various acts of gender create the idea of gender, and without those acts, there would be no gender at all. Gender is, thus, a construction that regularly conceals its genesis; the tacit collective agreement to perform, produce, and sustain discrete and polar genders as cultural fictions is obscured by the credibility of those productions – and the punishments that attend not agreeing to believe in them; the construction "compels" our belief in its necessity and naturalness. The historical possibilities materialized through various corporeal styles are nothing other than those punitively regulated cultural fictions alternately embodied and deflected under duress.

Consider that a sedimentation of gender norms produces the peculiar phenomenon of a "natural sex" or a "real woman" or any number of prevalent and compelling social fictions, and that this is a sedimentation that over time has produced a set of corporeal styles which, in reified form, appear as the natural configuration of bodies into sexes existing in a binary relation to one another. If these styles are enacted, and if they produce the coherent gendered subjects who pose as their originators, what kind of performance might reveal this ostensible "cause" to be an "effect"?

In what senses, then, is gender an act? As in other ritual social dramas, the action of gender requires a performance that is *repeated*. This repetition is at once a reenactment and reexperiencing of a set of meanings already socially established; and it is the mundane and ritualized form of their legitimation.[1] Although there are individual bodies that enact these significations by becoming stylized into gendered modes, this "action" is a public action. There are temporal and collective dimensions to these actions, and their public character is not inconsequential; indeed, the performance is effected with the strategic aim of maintaining gender within its binary frame – an aim that cannot be attributed to a subject, but, rather, must be understood to found and consolidate the subject.

Gender ought not to be construed as a stable identity or locus of agency from which various acts follows; rather, gender is an identity tenuously constituted in time, instituted in an exterior space through a *stylized repetition of acts*. The effect of gender is produced through the stylization of the body and, hence, must be understood as the mundane way in which bodily gestures, movements, and styles of various kinds constitute the illusion of an abiding gendered self. This formulation moves the conception of gender off the ground of a substantial model of identity to one that requires a conception of gender as a constituted *social temporality*. Significantly, if gender is instituted through acts which are internally discontinuous, then the *appearance of substance* is precisely that, a constructed identity, a performative accomplishment which the mundane social audience, including the actors themselves, come to believe and to perform in the mode of belief. Gender is also a norm that can never be fully internalized; "the internal" is a surface signification, and gender norms are finally phantasmic, impossible to embody. If the ground of gender identity is the stylized repetition of acts through time and not a seemingly seamless identity, then the spatial metaphor of a "ground" will be displaced and revealed as a stylized configuration, indeed, a gendered corporealization of time. The abiding gendered self will then be shown to be structured by repeated acts that seek to approximate the ideal of a substantial ground of identity, but which, in their occasional *dis*continuity, reveal the temporal and contingent groundlessness of this "ground." The possibilities of gender transformation are to be found precisely in the arbitrary relation between such acts, in the possibility of a failure to repeat, a de-formity, or a parodic repetition that exposes the phantasmatic effect of abiding identity as a politically tenuous construction.

If gender attributes, however, are not expressive but performative, then these attributes effectively constitute the identity they are said to express or reveal. The distinction between expression and performativeness is crucial. If gender attributes and acts, the various ways in which a body shows or produces its cultural signification, are performative, then there is no preexisting identity by which an act or attribute might be measured; there would be no true or false, real or distorted acts of gender, and the postulation of a true gender identity would be revealed as a regulatory fiction. That gender reality is created through sustained social performances means that the very notions of an essential sex and a true or abiding masculinity or femininity are also constituted as part of the strategy that conceals gender's performative character and the performative possibilities for proliferating gender configurations outside the restricting frames of masculinist domination and compulsory heterosexuality.

Genders can be neither true nor false, neither real nor apparent, neither original nor derived. As credible bearers of those attributes, however, genders can also be rendered thoroughly and radically *incredible*.

1. See Victor Turner, "Dramas, Fields and Metaphors" (Ithaca: Cornell University Press, 1974). See also Clifford Geertz, "Blurred Genres: The Refiguration of Thought," in *Local Knowledge, Further Essays in Interpretive Anthropology* (New York: Basic Books, 1983).

VIDEODROME
by **Scott Bukatman**

In *Videodrome*, which might stand as the ultimate statement on the place of the image in terminal culture, Cronenberg's overt fascination with McLuhanism is supplemented by what seems to be a prescient figuration of Baudrillard. The mediation of the image as a hyperlanguage and hyperreality allows Cronenberg to situate his bodily figurations and demands a reading through the tropes of postmodernism, in which the negation of a mind/body dichotomy takes its place within a set of such negated oppositions and boundary dissolutions, including self/other, private/ public, and spectacle/reality.

Videodrome presents a destabilized reality in which image, reality, hallucination, and psychosis become indissolubly melded, in what is certainly the most estranging portrayal of image addiction and viral invasion since Burroughs. "Videodrome" itself, apparently a clandestine television broadcast, is referred to as a "scum show" by its own programmers and depicts brutal torture and sadism in a grotesque display which exerts a strong influence upon its viewers. Cable station operator Max Renn desires "Videodrome": as a businessman he needs it to rescue his foundering station; as an individual he finds himself irresistibly drawn to its horrors. Renn must track down the source of this mysterious program (which emanates from either South America or Pittsburgh). Larger themes are connected to Renn's quest, such as the pervasiveness of the media-dominated spectacle in a postmodern world and, further, the passage beyond mere spectacle to the ultimate dissolution of all the boundaries which might serve to separate and guarantee definitions of "spectacle," "subject," and "reality" itself.

At times *Videodrome* seems to be a film which hypostatizes Baudrillard's most outrageous propositions. Here, for example, with remarkable syntactic similarity, Baudrillard and a character from Cronenberg's film are both intent upon the usurpation of the real by its own representation; upon the imbrication of the real, the technologized, and the simulated. The language is hypertechnologized but antirational; Moebius-like in its evocation of a dissolute, spectacular reality:

Jean Baudrillard:
"The era of hyperreality now begins...it signifies as a whole the passage into orbit,
as orbital and environmental model, of our private sphere itself.
It is no longer a scene where the dramatic interiority of the subject,
engaged as with its image, is played out.
We are here at the controls of a micro-satellite, in orbit,
living no longer as an actor or dramaturge but as a terminal of multiple networks.
Television is still the most direct prefiguration of this.
But today it is the very space of habitation that is conceived as both receiver
and distributor, as the space of both reception and operations, the control screen and terminal
which as such may be endowed with telematic power."[1]

Professor Brian O'Blivion:

"The battle for the mind of North America will be fought in the video arena – the Videodrome.
The television screen is the retina of the mind's eye.
Therefore the television screen is part of the physical structure of the brain.
Therefore whatever appears on the television screen emerges as raw experience for those who watch it.
Therefore television is reality and reality is less than television."

Cronenberg and Baudrillard both, in fact, seem to be following Debord's program that "When *analyzing* the spectacle one speaks, to some extent, the language of the spectacular itself in the sense that one moves through the methodological terrain of the very society which expresses itself in spectacle" (Thesis 11: the dictum strongly informs all the science fiction analyzed in this chapter). Baudrillard embraces a high-tech, alienating, and alienated science fictional rhetoric to explore high-tech alienation, while Cronenberg's horror films about the failure of interpersonal communications are an integral part of an industry which privileges the spectacular over the intimate, and pseudo-satisfaction over genuine comprehension. Both construct discourses of antirationalism in an attempt to expose and ridicule any process or history of enlightenment which might occur through the exercise of a "pure" reason. The complexity and evasiveness of Baudrillard's prose complements the visceral and hallucinatory image-systems of Cronenberg's cinema.

Videodrome presents a most literal depiction of image addiction. The title of the film is presented as a video image; following a flash of distortion, the title is replaced by another, this one on a diegetic television screen, while an accompanying voice-over announces, "CIVIC TV...the one you take to bed with you." Dr. Brian O'Blivion is the founder of the Cathode Ray Mission, a kind of TV soup kitchen for a derelict population. Scanning the rows of cubicles, each containing a vagrant and a television, Max Renn asks, "You really think a few doses of TV are gonna help them?" O'Blivion's daughter replies, "Watching TV will patch them back into the world's mixing board." On the street a derelict stands with his television set and a dish for change in what is presumably a watch-and-pay arrangement.

Within the diegesis, television frequently serves as a medium of direct address. Renn awakens to a videotaped message recorded by his assistant. O'Blivion refuses to appear on television "except *on* television": his image appears on a monitor placed beside the program's host (in a gesture reminiscent of Debord's own prerecorded lectures).[2] As Renn awaits his own talk-show appearance, he chats with Nicki Brand, the woman next to him, but an interposed monitor blocks any view of her. The image on the monitor is coextensive with its own background, however – Magritte-like – and consequently, the conversation is between a live Renn and a video Brand. Further examples of direct address proliferate, offering a preliminary blurring of any distinction between real and televisual experience.

This parody of McLuhanism serves as a backdrop to the enigma of *Videodrome*, which is finally revealed to be a government project. The explanation for *Videodrome* is at least as coherent as any from Burroughs: Spectacular Optical, a firm which specializes in defense contracts ("We make inexpensive eyeglasses for the Third World, and missile guidance systems for NATO"), has developed a signal which induces a tumor in the viewer. This tumor causes hallucinations which can be recorded, then revised, then fed back to the viewer: in effect, the individual is reprogrammed to serve the controller's ends. Burroughs, at his most paranoid, offered a similar vision of the subject: you are a programmed tape recorder set to record and play back who programs you who decides what tapes play back in present time.[3]

While the images which accompany the transmission of the *Videodrome* signal are not directly significant, it is the violence and sadism of *Videodrome* (the program) which "open receptors in the brain which allows the signal to sink in."

But as Barry Convex of Spectacular Optical asks Renn, "Why would anyone watch a scum show like *Videodrome*?" "Business reasons," is Renn's fast response, to which Convex retorts with a simple, "Sure. What about the other reasons?" Convex is correct: Renn's interest in the *Videodrome* broadcast transcends the commercial. "You can't take your eyes off it," is only his initial response in what becomes an escalating obsession. Asking what sort of program he might himself produce, a client asks, "Would you do *Videodrome*?" Coincident with his exposure to the *Videodrome* signal is his introduction to Nicki Brand, an outspoken, alluring radio personality for C-RAM radio.[4] Transgression thus functions in Renn's life in at least three modes: the social transgression represented by his soft-porn, hard-violence cable TV station; the sexual transgressions of his forays into sadomasochistic sexuality with Brand; and the political and sexual transgressions of *Videodrome's* sadistic presentations of torture and punishment. The three levels are linked in a spiraling escalation which culminates in Renn's own appearance on *Videodrome*, whipping, first Brand, then her image on a television monitor. Brand is the guide who leads Renn towards his final destiny; after her death, her image remains to spur him on. Her masochism might indicate a quest for real sensation: this media figure admits that, "We live in overstimulated times. We crave stimulation for its own sake." Brand wants to "audition" for *Videodrome*: "I was made for that show," she brags, but it might be more accurate to say that she was made by that show. Bianca O'Blivion tells Renn, "They used her image to seduce you."

The *Videodrome* program is explicitly linked by both Renn and Convex to male sexual response (something "tough" rather than "soft") and penetration (something that will "break through"). Renn takes on the "tough" sadistic role with Brand, and yet there is no doubt that it is she who controls the relationship, she who dominates. Similarly, the power granted by the *Videodrome* program to observe and relish the experience of torture and vicious brutality disguises the actual function of the program to increase social control: to establish a new means of dominance over the population. Renn is superficially the master of Brand and Videodrome, but ultimately the master becomes the slave. In a Baudrillardian revision of the Frankenstein myth, even Brian O'Blivion is condemned: *Videodrome's* creator is its first victim.

There is a distinctly Third World flavor to the mise-en-scène of the *Videodrome* program in its low-technology setting, electrified clay walls, and the neo-stormtrooper guise of the torturers. All this exists in contrast to the *Videodrome* technology: electronic and invisible, disseminated "painlessly" through the mass media. "In Central America," Renn tells Brand, "making underground videos is a subversive act." In North America it is too, it would seem, as the *Videodrome* signal subverts experience, reality, and the very existence of the subject.

Once again, it is the voluntarism of the television experience, the "free choice" of the viewer, which permits the incursion of controlling forces. A strictly political-economic reading of *Videodrome* could easily situate the work within Debord's *Society of the Spectacle*. Images stand in for a lost social whole, the spectator's alienation is masked via the reified whole of the spectacle, the capitalist forces are thereby able to reproduce themselves at the expense of the worker/consumer/spectator. Cronenberg has replaced the structures of power absent from McLuhan's schema: Brian O'Blivion envisioned *Videodrome* as the next step in the evolution of man, but his utopian technologism is usurped by the technocratic order of state control.

But *Videodrome* moves beyond this classically political reading through its relentless physicality. The film's politics have less to do with economic control than with the uncontrolled immixture of simulation and reality. In *Videodrome* the body literally opens up – the stomach develops a massive, vaginal slit – to accommodate the new videocassette "program." Image addiction reduces the subject to the status of a videotape player/recorder; the human body becomes a part of the massive system of reproductive technology (you are a programmed tape recorder). The sexual implications of the imagery are thus significant and not at all gratuitous: video becomes visceral.[5]

Following his own exposure to the *Videodrome* signal, Renn begins his series of hallucinations with a spectacular immersion in the world of the spectacle. When his visiting assistant, Bridey, reaches for the *Videodrome* cassette, Renn assaults her. In a series of shot/reverse shot pairings, Bridey becomes Brand, then Bridey again. Disoriented, Max apologizes for hitting her. "Hit me?" she answers, "Max...you didn't hit me." The videotape she has delivered from Brian O'Blivion breathes and undulates in his hand; he drops it and kicks it lightly, but it only lies there, inert. As O'Blivion tells him: "Your reality is already half-video hallucination."

The videotaped message from O'Blivion suddenly becomes more *interactive*. "Max," he says, all trace of electronic filtering abruptly gone from his voice. "I'm so happy you came to me." O'Blivion explains the history of the *Videodrome* phenomenon while being readied for execution: the executioner is Nicki Brand. "I want you, Max," she breathes. "Come to me. Come to Nicki." Her lips fill the screen, and the set begins to pulsate, to breathe. Veins ripple the hardwood cabinet; a videogame joystick waggles obscenely. All boundaries are removed as the diegetic frame of the TV screen vanishes from view: the lips now fill the movie screen in a vast close-up. Renn approaches the set as the screen bulges outward to meet his touch, in a movement which literalizes the notion of the screen as breast. His face sinks in, his hands fondle the panels and knobs of the set as the lips continue their panting invitation.

Cronenberg moves the viewer in and out of Renn's hallucination by creating a deep ambiguity regarding the status of the image. It is easy to accept the attack on the assistant as real, although the transmigration of identities clearly demarcates Renn's demented subjectivity. Yet, it turns out, the attack was entirely hallucinated: the "real" cinematic image is unreliable. In the extended hallucination of the eroticized, visceral television, the filmmaker gracefully dissolves the bounds that contain the spectacle. O'Blivion's voice is no longer marked as a mediated communication once the electronic tone of his speech ceases. The TV screen is contained by its own frame, but Cronenberg's close-up permits the image to burst its boundaries and expand to the nondiegetic limits of the cinema screen. In a later hallucination, a video-Brand circles Renn with whip in hand, profferring it for him to wield. The image moves from video hallucination to cinematic reality within a single shot: Renn accepts the whip, but Brand is now no longer present in corporeal form; she only exists, shackled, on a TV screen. Renn attacks the bound(ed) image with fervor: another moment which recalls the visual punning of René Magritte. These shifts in visual register mark the passage from spectacle as visual phenomenon to spectacle as new reality.

Cronenberg, then, does not mythologize the cinematic signifier as "real," but continually confuses the real with the image and the image with the hallucination. When Renn pops a videotape into his machine, Cronenberg inserts a blip of video distortion over the entire visual field before cutting to O'Blivion's image on-screen. This does not mark the hallucination, but it "infects" the viewer with an analogous experience of dissolution and decayed boundaries. These confusions, between reality, image, and hallucination, pervade the film. There is no difference in the cinematic techniques employed, no "rational" textual system, which might serve to distinguish reality from hallucination for the film viewer. Each moment is presented as "real," that is, as corresponding to the conventions of realist filmmaking. Discourse itself is placed in question in *Videodrome* through the estrangement of cinematic language. Where the hallucination might have begun or ended remains ambiguous, uncertain. These unbounded hallucinations jeopardize the very status of the image: we must believe everything and nothing, equally. In the words of the Master Assassin Hassan-i-Sabbah, (used as the epigraph to Cronenberg's adaptation of *Naked Lunch* [1991]) *"Nothing is true. Everything is permitted."*[6]

Renn hallucinates his appearance on *Videodrome*, but is *Videodrome* a program comprised entirely of recorded hallucinations? If so, then there is a progression from hallucination, through image, to reality: the scene is real because it is televised, it is televised because it is recorded, it is recorded because it is hallucinated. The illusory rationality which guides the society of the spectacle emanates from the irrational recesses of the libidinal mind. On the medium of television, Baudrillard writes: "The medium itself is no longer identifiable as such, and the merging of the medium and the message (McLuhan) is the first great formula of this new age. There is no longer any medium in the literal sense: it is now intangible, diffuse and diffracted in the real, and it can no longer be said that the latter is distorted by it."[7] Society becomes the mirror of television (Kroker) as television becomes the new reality. A slit opens in Max Renn's stomach, and Barry Convex holds up a videocassette, which breathes. "I've got something I want to play for you," he says and inserts the cassette. The human body thus becomes a part of the technologies of reproduction observed by Jameson: the ultimate colonization under late capitalism and the ultimate penetration of technology into subjectivity. The reprogrammed Renn later retrieves a gun from this new organ, a gun which extends cables and spikes into his arm in an inversion of McLuhan's sense of technology as human extension. Here the man becomes the extension of the weapon: a servo-mechanism or perhaps a terminal. There is none of McLuhan's hypothetical "numbing" in this most painful of cinematic displays. We are instead still trapped within a universe which seems to be someone else's insides.

In its themes and structure the film serves as a graphic example of what Baudrillard termed "the dissolution of TV into life, the dissolution of life into TV." Baudrillard terms this immixture *"viral,"* echoing Burroughs's injunction that "image *is* virus." The viral metaphor is strikingly apt when applied to *Videodrome*: the literalized invasion of the body by the image and the production of tumors which produce images. Image is virus; virus virulently replicates itself; the subject is finished.

Body and image become one: a dissolution of real [sic] and representation, certainly, but also of the boundaries between internal and external, as the interiorized hallucination becomes the public spectacle of the *Videodrome* program. Here *Videodrome* echoes *The Simulacra,* in which a character's psychosis results in a physical transformation (his organs telekinetically appear outside, as objects in the room are reciprocally introjected). In the postspectacle society delineated by Baudrillard, all such boundaries will dissolve, will become irrelevant through the imperatives of the model of communication (simultaneous transmission and reception):

> *In any case, we will have to suffer this new state of things,*
> *this forced extroversion of all interiority, this forced injection of all exteriority*
> *that the categorical imperative of communication literally signifies;...*
> *we are now in a new form of schizophrenia.*
> *No more hysteria, no more projective paranoia, properly speaking,*
> *but this state of terror proper to the schizophrenic: too great a proximity of everything,*
> *the unclean promiscuity of everything which touches,*
> *invests and penetrates without resistance.*[8]

The subject has "no halo of private protection, *not even his own body,* to protect him anymore" (emphasis mine). The works of David Cronenberg, as well as those of Philip Dick, repeat several of these tropes. The subject is in crisis, its hegemony threatened by centralized structures of control, by a technology which simultaneously alienates and masks alienation, by a perception of its own helplessness. Even the last retreat, the physical body, has lost its privileged status: hence the schizophrenic terror undergone by the protagonists. Even the libido, site of the irrational, seat of desire, is invaded, enlisted in the furtherance of an obsolescent technological rationalism.

Again, these texts are not simply reactionary moments of nostalgia, but bring a profound and progressive ambivalence to the imbrication of simulation and reality, subject and other. The slippage of reality that marks the textual operations of *Videodrome* can certainly be associated with the commensurate process in the writings of the saboteur Burroughs, who repeatedly declared that we must "Storm the Reality Studio and retake the universe."[9] This cinematic metaphor reaches a kind of apotheosis in *Videodrome,* as the images flicker and fall, their authority ultimately denied, but there is no glimpse of a Reality Studio behind the myriad levels of reality production.

1. Jean Baudrillard, *The Ecstasy of Communication,* Foreign Agents Series, ed. Jim Fleming and Sylvere Lotringer (New York: Semiotext(e), 1988): 128.

2. See Guy Debord, "Perspectives for Conscious Alterations in Everyday Life," in *Situationist International Anthology,* ed. Ken Knabb (Berkeley, Calif.: Bureau of Public Secrets, 1981), for a transcript of one of these performances.

3. William S. Burroughs, *The Ticket That Exploded* (New York: Grove Press, 1967): 213.

4. The acronym suggests the computer term, Random Access Memory.

5. For an important analysis of the figuration of the body in *Videodrome,* from an explicitly feminist theoretical position, see Tania Modleski, "The Terror of Pleasure: The Contemporary Horror Film and Postmodern Theory," in *Studies in Entertainment: Critical Approaches to Mass Culture,* ed. Tania Moledski (Bloomington: Indiana University Press, 1986).

6. After detailing the similarities between the work of Burroughs and Cronenberg, I must admit that *Naked Lunch* reveals their profound differences; most notably in the area of sexuality.

7. Jean Baudrillard, *Simulations,* Foreign Agents Series, Trans. Paul Foss, Paul Patton, and Philip Beitchman (New York: Semiotext(e), 1983): 54.

8. Baudrillard, *Ecstasy*: 132.

9. William S. Burroughs, *The Soft Machine* (New York: Grove Press, 1966): 155.

THE XEROX DEGREE OF VIOLENCE

by **Jean Baudrillard**

Hatred: instead of deploring the resurgence of an atavistic violence, we should see that it is our modernity itself, our hypermodernity that generates this type of violence and these special effects, among which terrorism plays its part as well. Traditional violence is much more enthusiastic and sacrificial. Ours is a simulated violence in the sense that, more than being based in passion and instinct, it springs up from the screen. In a certain way, violence has a potential existence in the screen and in the media which pretend to record and broadcast it but which, in reality, come before it as well as appeal to it. Here, like everywhere else, the media precedes this violence (just as it precedes terrorist acts). This is precisely what makes it a specifically modern form, why it is impossible to assign genuine causes to it (political, sociological or psychological). One feels that these types of explanations are failing. Following the same idea, it is rather meaningless to blame the media for propagating violence through broadcasting and sensationalism. For the screen, this virtual surface, protects us rather efficiently – no matter what we usually think – from the actual content of the pictures. Due to the continuous aspect of the screen, the spectacle of violence does not lead to violent behavior. The threat which we cannot ignore is the violence of the media itself, the violence of virtuality and its nonspectacular proliferation. What is to be feared is not the psychological progression but the technological progression of violence; a transparent violence, the type of violence that leads to the disembodiment of reality and of all systems of reference. It is the Xerox degree of violence.

It is because our society has evolved to a point where historical and class violence are no longer given space that it generates a virtual and reactive violence. A sort of hysterical violence – the same way we talk about a hysterical pregnancy – and which, like the latter, gives birth to absolutely nothing. The same with hatred, which could also be seen as an archaic urge, but, paradoxically, because it is disconnected from its object and its aims, is in fact contemporary with the hyperreality of our large metropolis. One can identify a primary form of violence: the violence that relates to aggression, oppression, violation, to power struggles, humiliation, and despoliation. The unilateral violence of the strongest. To which one can reply with a contradictory violence: historical and critical violence, violence of the negative, of the rupture, transgressive violence (to which can be associated the violence of analysis and interpretation). All of these are definite forms of violence, with an origin and an end. Forms of violence for which one can identify the causes and effects and which correspond to a transcendence, that of power, history or meaning.

In opposition to this, a strictly contemporary form of violence is developing, more subtle than sheer aggression: the violence of dissuasion, pacification, neutralization, and of control (the violence of smooth extermination, genetic violence, the violence of communication). The violence of consensus and conviviality that tends to abolish through drugs, prophylaxis and by psychic and media regulation, the very roots of evil and hence, all types of radical thought. The violence of a system that tracks down all forms of negativity and peculiarity (including the ultimate form of peculiarity: death itself). The violence of a society which bans any negativity and conflict, which bans us from dying. A violence that abolishes itself, and therefore cannot be answered by an equal form of violence except through hatred.

Born from indifference, and irradiated by the media, hatred is a *cool*[1], discontinuous form that focuses on some object or other. Hatred acts without conviction, without warmth, it wears itself out through *acting out* and also, as can be seen in recent episodes of suburban delinquency, through the images of hate and their immediate repercussions. If traditional violence was an answer to oppression and conflict; hatred, for its part, is an answer to consensus and conviviality. Our eclectic culture encourages the blend of opposite tendencies and the coexistence of all differences within a great cultural melting pot. But let us not be mistaken: it is precisely this multicultural atmosphere, this tolerance and synergy that stirs up the temptation of a global abreaction, of a visceral repulsion. Synergy provokes allergy. Too much protection compromises the immune system. The antibodies are laid off and eventually turn against the organism. Hatred belongs to this category. Like most modern diseases, it derives from auto-aggression and from an auto-immune pathology. We are not ready to endure the artificial immunity that has developed within our metropolis. We behave like a species which has eliminated all its natural predators, and which is rapidly doomed to disappearance or auto-destruction. We use hatred to protect ourselves against this weakness in the other, against the enemy, against adversity. This type of hatred, that thrives within a sort of artificial adversity, is aimless. Thus, hatred is a kind of fatal strategy against the pacification of life. In its ambiguity, hatred is a desperate claim against the indifference of our world and is in all likelihood, a mode of relating much more powerful than consensus or conviviality.

The recent shift from violence to hatred characterizes the shift from an object-passion to a passion without object. Pure and undifferentiated violence; in a certain way, violence in the third degree. Contemporaneous with the exponential violence exemplified in terrorism and in all the viral and epidemic forms of contagion and chain reaction. Hatred is more unreal, more elusive in its manifestations than simple violence. One can see this clearly with racism and delinquency. This is why it is so difficult to prevent them, no matter how repressive the methods used against them. Neither can hatred be demotivated nor demobilized since it doesn't have any explicit motive or motivation. It can hardly be punished since, most of the time, it turns on itself. Hatred is the epitome of a passion battling with itself.

We are doomed to the replication of the same patterns in a never-ending identification, in a universal culture of identity, and from there derives an immense resentment: hatred of self. Not hatred of the other – to which a superficial interpretation of racism would lead – but of the loss of the other and of the resentment over this loss. One usually assumes that hatred is aimed at the other, hence the illusion of being able to blunt it through tolerance and a respect for differences. However, instead of being a rejection of the other, hatred (racism, etc.) would be much closer to what we could call *a fanaticism of otherness.* It desperately attempts to make up for the loss of the other through the exorcism of an artificial other which could be anyone. In a lobotomized world where conflicts are immediately contained, it aims at reawakening otherness, if only to destroy it. It attempts to escape from this fatal identification, from this autistic confinement to which we are doomed by the very evolution of our culture. A culture of resentment in which, behind the resentment towards the other, hides the resentment towards self, towards the dictatorship of self, a resentment that finds its paroxysm in self-destruction.

A crepuscular passion, this is how we should consider hatred, in all its ambiguity. And at the same time as symptom and cause of this brutal loss. Loss of social interactions, of otherness, loss of conflicts and eventually loss of the system itself, doomed to gravitational collapse. Symptom of the end or symptom of the failure of modernity, if not the end of History, for paradoxically, History has never had an end since no one has ever solved the problems it created. It would be more appropriate to say that we have reached a point beyond the end, although nothing has truly been solved. And in today's hatred, there is a resentment towards everything that has never happened. Hence the urge to rush things in order to escape from the system, to move onto something else, to contribute to the birth of something new: to events that come from elsewhere. In this *cool* fanaticism comes a millenarian form of provocation.

We all possess hatred. We cannot help having it. We feel an ambiguous nostalgia for the end of the world, that is, to give it an end, a purpose, no matter the price; even if it entails resentment and a total rejection of the world.

October 1995

Translated from the French by Thierry Duval

1. In English in the original text.

BLUE DATA[1]
by **Lia Gangitano**

Upon David Cronenberg's entry into the virtually constructed world of computer-generated spaces in his forthcoming film, *eXistenZ*, the occasion arises to review the director's early aesthetic foothold in the tangible realities of everyday horror, for example, as embodied in the concrete and glass of modern architecture. His tendency toward formal restraint is evident in the operating theater of *Dead Ringers* (1988) and the austere gray freeways of *Crash* (1996), but the origins of this controlled aesthetic (and its incumbent rupture), as aligned with Toronto's icy institutional backdrop, are to be found in the films that predate his first commercial release, *Shivers* (1975). Cronenberg's development and return to certain ideas that are foreshadowed in these unencumbered experimental films have formed a basis for this investigation into the relationship between his films and contemporary artistic practice.[2]

The impulse to center an exhibition of contemporary art on the work of a prolific and increasingly mainstream director naturally departs from an investigation of some of his more obscure proclivities and would seek to excavate his early status as an artist of radical character. Reviews for his earliest films *Stereo* (1969) and *Crimes of the Future* (1970) align Cronenberg with figures ranging from Godard to Burroughs to Warhol[3], while *Shivers* garnered success through an affiliation with the exploitation movie. This could be read as a sign of the times, illustrating the transition from aesthetic to more popular goals, but this is an oversimplification in that even Cronenberg's particular brand of brainy shock demonstrates a transgressiveness of content and form that tends to be distinguished as art. The current tendency of contemporary art to aspire to other media and modes of production comprises another aspect of a reciprocal relationship that forms a starting point for the exhibition *Spectacular Optical*.

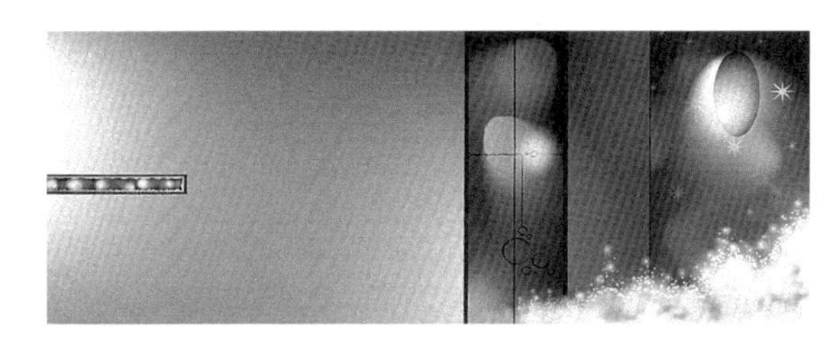

Fig. 1

Because of the relentless austerity and institutional bleakness of Cronenberg's early films, it is tempting, given the nature of this project, to trace his development in terms of the artistic precedents that coincided with their production, namely Minimalism and Conceptual art: "...reductivist tendencies formed a primary fascination in the 60s because they resulted in the production of austere and mechanistic objects whose aggressive passivity and stylish indifference toward the spectator evoked feelings of alienation and dehumanization reminiscent of those experienced in everyday life."[4] This is not to disregard the context of experimental film being produced in Canada and elsewhere, and "the Ontario underground scene," of which Cronenberg became an important part. After leaving his university studies in biochemistry in favor of the English department, it was apparently seeing the student film, *Winter Kept Us Warm* (1965) by David Secter, that encouraged Cronenberg to make his first film. Without an existing film department at the University of Toronto, Cronenberg's intellectual investment in literature, philosophy and writing strongly influenced his self-taught approach to filmmaking.

He completed two films during his undergraduate years, including *Transfer* (1966), his first black and white 16mm film, that depicts a self-exiled psychoanalyst pursued by a dependent patient to his refuge in a barren field. Their dialogue is fraught with sexual innuendo and culminates with a therapy session held out in the snow. *From the Drain* (1967), his next short, finds two men cruising each other in a bathtub. They hint at being survivors of some ambiguous biological war. Their awkward flirtation is fueled by the threat of something lurking in the drain. Eventually a stop motion, weedlike thing emerges, strangling one of them. I mention these films because of their uncharacteristic focus on dialogue, something that is notably absent in Cronenberg's subsequent two films, and certainly subsidiary to other means of propelling the plot in his better known early commercial films, *Shivers*, *Rabid* (1976), and *The Brood* (1979).

Fig. 2

Fig. 3

Although filmed in black and white, everything remains gray – the images, their implications, even their purpose.[5]

Viewing *Stereo* is a peculiar experience, something analogous to torture by minimalism (admittedly, my favorite kind). A remoteness of location is established by the opening shot, in which the main character arrives via helicopter to join an experiment in telepathy being conducted at the Canadian Academy for Erotic Inquiry. The film is set in architect John Andrews' Scarborough College, "a building of such overpowering antiseptic grandeur that it steals the film from the human performers."[6] Cronenberg's choice of location sets a precedent for future works, in which rational institutional structures are placed in stark contrast with the irrational projects that infect them. It has been said of Scarborough College that: "Within an overall conceptual framework the design evolves in a rational and systematic manner.... The expression of the building is pursued as a logical extension of the program. [Andrews'] strength here lies in an ability to manipulate forms into compatible orders, but, he claims, 'only the forms that start to want to happen.' While Louis

Kahn's 'let a building be what it wants to be' is a basic tenet of his work...there is a sense of inevitability in Andrews' architecture that is absent in Kahn's more formalized solutions."[7] The notions of inevitability and evolution, invoked by the building, are heightened by Cronenberg's particular orientation toward a precarious future, in which science has become corrupted by the eccentric desires of an overly creative genius.

Early in his career, Cronenberg established his approach to these states of evolved crises in the not-too-distant future, not through an adherence to the available codes of science fiction, but by maximizing the extreme attributes of modernity. The fantastical systems deployed by science fiction seem incompatible with the rational framework to which Cronenberg consistently refers, if only to fracture, in his films. Also absent is any retreat to the past or toward supernatural forces, typical of the horror genre. He implies that the ultra-rational thought structures that would spawn such constructions as the Canadian Academy for Erotic Inquiry or subsequent office parks, apartment complexes and scientific institutions are inherently anxiety-producing in their oppressive relationship to the adaptable yet vulnerable body.[8] One critic has noted: "The repeated spatializing of the body results in a treatment of space very different from more traditional horror filmmakers. Cronenberg's films contain very little sense of the space off-screen, that realm of the unknown and taboo which has served as the central spatial metaphor for the genre.... The spaces are relentlessly interior – apartments, offices, laboratories.... Enclosure, separation, and isolation define the on-screen realm of Cronenberg's filmmaking, and the profusion of slow-tracking shots across these spaces emphasizes the emptiness of what is present, rather than the completion demanded by what is not. In other words, the forces of the repressed are always already present and are never displaced to some 'other,' absent region to await reemergence."[9] In reference to his turning down directing *Interview with the Vampire*, Cronenberg has commented: "I'd been told it was a modern rethinking of vampirism. [...] But when I started to read it, I didn't find it very modern. I found it very gothic and very old-fashioned and very florid." Instead, the vampiric qualities of the modern era are evoked through a calculated formal approach, often divulged though his treatment of space. Much like the poetics of exploded repression described by Canadian photographer, Jeff Wall, in reference to Dan Graham's *Alterations Project*,[10] Cronenberg and the artists in *Spectacular Optical* address the horror show of Modernism by responding to its insidious effects, as evidenced by the structures of everyday life.

The vampire is neither alive nor dead, but exists in an accursed state of irremediable tension and anxiety. Although his symbolic identity is complex and goes beyond its function in this analysis, he embodies a certain sense of cosmic grief which is a diffracted image of a concrete historical uneasiness. ...[T]he vampire signifies not simply the unwillingness of the old regime to die, but the fear that the new order has unwittingly inherited something corrupted and evil from the old, and is in the process of unconsciously engineering itself around an evil centre. This presence of the phantasm of the vampire in the modern, liberal consciousness signifies an unresolved crisis in the creation of the modern era itself. Thus the vampiric symbolism persists as a codified form of expression of unease regarding the inner structure of the modern social order and corresponding psyche, particularly its commitment to calculation and rationality.... Vampiric symbolism is a disturbance in the historical process of construction of theoretical beings – abstract citizens – through technique, planning, contracts, and 'value-free' calculation. The Vampirism is thus the 'inner speech' of that being – the ruler, the caesar, the prince – whose theoretical invisibility is constructed as the great task of tragic modern architecture, the undead art.[11]

Fig. 4

Fig. 5

The only remnants of Cronenberg's more ornate historical inclinations are the somewhat Shakespearean costumes worn by the actors in *Stereo*. The film is silent except for intermittent voice-over, providing the observations of unseen researchers as well as describing the theories of mastermind parapsychologist, Luther Stringfellow. Cronenberg's treatment of the human subjects (echoed in his recent film, *Crash*) is described in terms of literal objectification: "...the camera plays over them as if they were sculpture. Cool, and slowly getting colder, in the final minutes each character seems like a piece of statuary in relief against John Andrews' grainy architecture."[12] This relationship between the captive body and oppressive architecture is further developed in Cronenberg's later films. In *Stereo*, the body is contained by and subjected to an overpowering structure, whereas in *Shivers*, the modern high-rise apartment building, Starliner Towers, actually takes on the characteristics of an infected body, with parasites travelling through the plumbing and ventilation ducts.[13]

The theme of addiction, initiated in his earliest films, continues to be of central importance. As well as acknowledging the potential of spaces to enact physical change on the bodies that inhabit them, Cronenberg elicits the evolutionary potential of substances and forces that, once introduced into the body, implicitly transform its makeup. This inevitability is portrayed as being consequential to normal conditions, insinuating the horror that one may try to reject is already latent in the human condition. In *Stereo*, "psychic addiction" leads one subject to drill a hole in his scull to relieve overwhelming telepathic data. This idea resurfaces quite literally in *Scanners* (1980), but also infuses Cronenberg's work as a whole. In reference to *Crash*, he has stated:

"Addiction, to me, is a very interesting phenomenon – especially biological addiction. Because there's a sense in which addiction merges with evolution – when you incorporate some other chemical process into your body, and it becomes a natural and necessary part of your body, then you've changed yourself – you've really transformed. I'm interested in transformation, and the extent to which evolution, in any species, involves a strange process whereby something is incorporated into your development as an animal that becomes necessary – which before was perhaps not even part of your metabolism – and now you're a different animal."[14] The evolution of the subjects of *Stereo* is willfully aided by the removal of certain areas of their brains as well as vocal cords, so that they communicate only by "telepathy heightened by sensory awareness, i.e., sexual stimulus."[15] The sexual stimulation is induced by aphrodisiacs. In one scene, the bottle reads, "Love Conquers All." Luther Stringfellow's project is ultimately irrational, driven by misguided utopian goals, and at odds with its brutal environment. His hope to replace the conventional family structure with the telepathic commune fails. Because one subject wants to be alone, she is plagued by "morbid telepathic images: disintegration, vampirism, necrophilia..." Two subjects commit suicide and the film concludes with a report on the injured subject's healing head wound.

Fig. 6

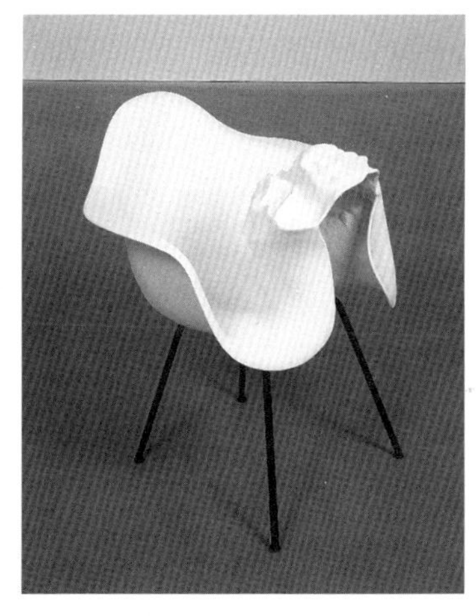

Fig. 7

Crimes of the Future takes place in a series of modern institutional settings, dubbed, in order of succession, The House of Skin, the Institute of Neo-Venereal Disease, and the Oceanic Podiatry Group, where our hero, Adrian Tripod (played by *Stereo's* leading actor, Ronald Mlodzik), slowly unravels the plot inadvertently set in motion by his former mentor, the mad dermatologist Antoine Rouge. The adult female population has been annihilated by Rouge's Malady, a disease that involves "severely pathological skin conditions induced by contemporary cosmetics." At The House of Skin, male survivors, while getting in touch with their femaleness through vague forms of crossdressing, are plagued by the emission of "pathological fluids." The excretions, labeled Rouge's foam, initially attract others to ingest them as aphrodisiacs, but later cause a severe case of v.d. Tripod visits his estranged colleague at the Institute to find him inflicted with the disease, apparently contracted from one of his patients. The effects of the disease include the development of "a galaxy of complex yet functionless organs." Although surgically removed and stored in specimen jars, Tripod's colleague feels compelled to steal them from the lab. After joining the Oceanic Podiatry program, Tripod is approached by a member of a conspiratorial organization of heterosexual paedophiles whose aim is to impregnate a girl who has avoided Rouge's Malady. She is kidnapped, but the film ends as Tripod hesitates – sensing the reincarnated presence of Antoine Rouge.

The imposing architectural environments of *Crimes of the Future* are made weirder by the use of an aquatic soundtrack, in addition to intermittent voice-over. Cronenberg has commented, "In *Crimes* I used a second soundtrack other than the voice-over, made up of deep-sea creatures, dolphins, shrimp. The sound of water is very present. In a sense the soundtrack was meant to be Darwin's voyage of the *Beagle*. I thought it was an underwater ballet. I wanted to create the feeling that you were watching aliens from another planet. There is a science-fiction element, but not as explicit as the genre demands."[16] Endless parallels between *Stereo* and *Crimes of the Future* and Cronenberg's subsequent films can be drawn. Recurring themes, such as the body's development of "creative cancers," are initiated in rudimentary form. Appendages developed in the patients of Hal Raglan (*The Brood*) bear a striking resemblance to the contents of the jelly jar props in *Crimes*. His distinctive use of architecture and institutional décor, with an indulgence in roving corridor shots, is firmly established in this set of early films. "Strikingly, *Dead Ringers*' own twin movies are Cronenberg's early featurettes *Stereo* and *Crimes of the Future*. In the aquamarine design of the Mantle twins' apartment, and their strange 'otherness', is the fishtank approach Cronenberg applied to his characters in those early avant-garde experiments. This is exactly signaled in the opening sequence, in which the young Mantles discuss the fact that humans are compelled to have sex only because they don't live in water."[17]

Cronenberg's early experimental films also provide a reference point for his continuing effort to define his work outside the confines of existing genres associated with independent as well as Hollywood filmmaking. As the limits of transgression are consistently normalized by such categories, aesthetic strategies that transform comfortable notions of mortality, sexuality, and moral value become more necessary. The horror created by the breakdown of existing structures, both ideological and formal, constitutes a defining characteristic of the cultural moment that *Spectacular Optical* reflects through contemporary artistic practice. Cronenberg has noted, "Subversion is essential to art...[and] if you are working within a genre, it's more simple to subvert. If you are not working within a genre, then it's a much more subtle thing. [...] One of the things about narratives is that they are predictable, comfortable things. The possibility for twists, turns, subversion are really quite endless. When you are inventing your own form...you don't have that possibility. The form itself is the subversive thing."[18]

Cronenberg's contribution, when recognized as such, is generally attributed to his bodily fixations: "...it was Cronenberg's resiting of horror from the realm of the gothic to the body that thrilled and shocked audiences."[19] He has commented: "There's a Latin quote that goes 'Timor mortis conturbat me,' which, roughly translated, means 'The fear of death disturbs me'. Death is the basis of all horror, and for me death is a very specific thing. It's very physical. That's where I become Cartesian. Descartes was obsessed with the schism between mind and body, and how one relates to the other. The phrase 'biological horror' – often attached to my work – really refers to the fact that my films are very body-conscious. They're very conscious of physical existence as a living organism, rather than other horror or science-fiction films which are very technologically oriented, or concerned with the supernatural, and in that sense are very disembodied."[20] The term "biological horror" was no doubt inspired by *Shivers* (also known by the titles *The Parasite Murders* and *They Came from Within*). This film sets out a new direction for low budget, yet commercial film-making for Cronenberg, and marks the beginning of his innovative use of special effects, in this case, low-tech methods involving balloons. However, the basic premise – the conflict between the rational and the irrational, with its implied link to addiction as an evolved human state – remains at the core of the film, which is played out within the confines of a modern apartment complex. The protagonist, Dr. Emil Hobbes, has developed a parasite whose purpose is to perform the function of human organs; but his true aim is to combat rationality. His human experiment is initiated by introducing a nymphomania-inducing parasite into the body of a schoolgirl who lives in Starliner Towers. Early in the film, Hobbes attempts to reverse his error by annihilating the

Fig. 8

Fig. 9

parasite, thus killing the girl, and then committing suicide. But the parasite has already spread throughout the building as well as its inhabitants. Like many of Cronenberg's creative scientists, Hobbes sets an insidious plague into motion. The opening sequence of the film is an infomercial-like slideshow, showcasing Starliner Towers, a planned living environment that fulfills all the domestic and lifestyle needs of its inhabitants under one roof. The self-contained, strictly controlled complex offers the ultimate in comfort, safety and convenience. Cronenberg's choice of the glassy high-rise as the site of this experiment-gone-wrong posits a critique of its quixotic, isolationist aims, and evokes a particular unease, as the chaos unleashed will spread exponentially from this point forward.

The combination of disorienting, mirroric invisibility with a monumentalism which is rigid, systematic, and empty of satisfying symbols of power and authority, makes the glass tower a disturbing phenomenon. Indeed, it becomes positively frightening at the point of realization that its symbolic system has triumphed over the ponderous monumentalism of earlier, European-style totalitarianism.

This implies that a new historical phase of oppression, more complex,
sophisticated and deeply embedded in the habitual and the unconscious,
has matured within capitalism. The system announces itself
in these neutral and functional, systematic feats of precision design and engineering.
They tend to signify the historical fact of redefinition of the ideal of rationality,
the historically complete enclosure of possibilities once hoped for from rationality,
as concretized in planning, by the institutions they glamourize. This effect,
of total enclosure of the realm of rationality by a monument to historical emptiness,
indicates that the symbolization processes generated by these buildings are irrational
in a new way. They express the emergence of a new stage of social irrationality
previously only latent in historical development.[21]

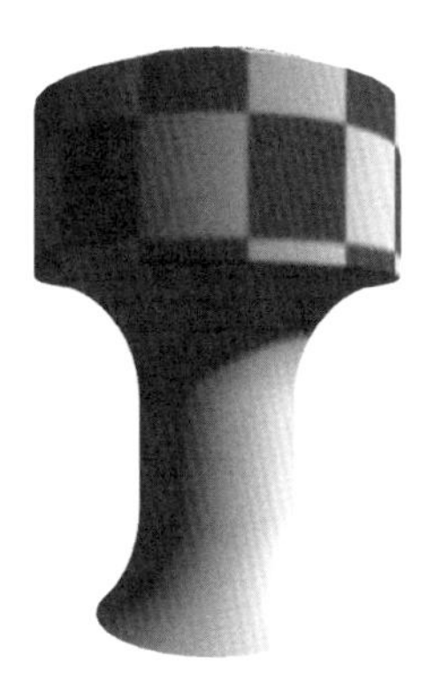

Fig. 10

Fig. 11

It is important to note, while tracing the trajectory of Cronenberg's known and unknown films in relation to contemporary art, that in 1976 he wrote and directed *The Italian Machine*, and in 1979 he co-wrote and directed *Fast Company*. These films focus exclusively on topics that surreptitiously sneak to the surface of most of his better-known films, and could be positioned as examples of his conspicuous fetishization of the machine, as well as his strange love for and sampling of things associated with so-called "low" culture. Somehow, it comes as little surprise that Cronenberg is into racing cars and motorcycles. *The Italian Machine*, a short film made for television, "...fuses the director's obsession with racing machinery as art..."[22] The conflict portrayed is between biker culture and the art world. A collector of modern art has purchased a rare Ducati Desmo Super Sport, which he displays in his living room as sculpture. The bikers learn of this offense and hatch a plan to get the bike back by posing as writers for *Techno Art World* magazine. Certain attitudes about art are disclosed in this film, in which the art collector, secluded in his modern home, embodies the distant, calculating individual, incapable of truly understanding the Ducati.

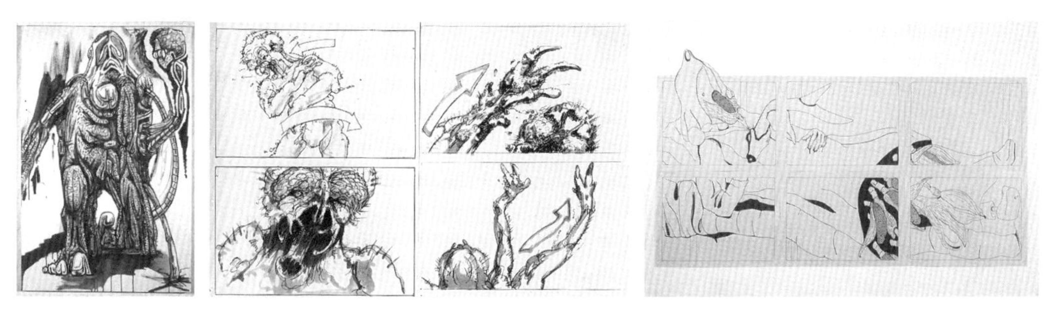

Fig. 12 **Fig. 13** **Fig. 14**

Cronenberg's attraction to fast cars, t&a, feathered hair, rock music, etc. is given full reign in *Fast Company*. His penchant for exploring interiors, those of both human and mechanical bodies, is manifested here in surgery-like scenes of drivers fixing their cars. Cronenberg notes: "What I like most about the movie came out of my appreciation and understanding of race cars and racing machinery, which I get very metaphysical and boring about.... I knew about racers and raced myself. I got the drag racers' particular version, which is very much a beer-drinking, wet T-shirt thing. They even had T-shirts that said 'Suck my pipes': a great phrase. I made sure I got that in. Great stuff. Not my sensibility, but definitely theirs. I was doing a bit of documentary filmmaking with that movie. I was reading those Hot Rod magazines and was ready to build myself a hot Camero."[23] These and other adolescent suburban fascinations loom, for example, in his portrayal of Rose and Hart as a biker couple in *Rabid*; or in his choice of school uniform for the first sex-crazed victim of *Shivers*; or in the character of the sleazy porn mogul of *Videodrome*, our hero, Max Renn.

"Unlike you, Max, it has a philosophy, and that is why it is dangerous."[24]

Fig. 15

Fig. 16

Fig. 17

With the marketing tools of the exploitation movie behind it, *Shivers* gained cult status for Cronenberg, as well as made him the subject of critique, particularly through feminist readings of his films. As with his next feature, *Rabid*, the specter of censorship appears only to solidify Cronenberg's resistance to political readings of his films. He has commented: "I would never censor myself. To censor myself, to censor my fantasies, to censor my unconscious would devalue myself as a filmmaker. It's like telling a surrealist not to dream. The way I portray women is much more complex than any ideological approach is going to uncover."[25] Criticism of his representation of female sexuality in *Shivers* as too passive and in *Rabid* as too predatory, provides the opportunity for his dismissive "you just can't win" response. He is equally dismissive of metaphorical readings of his films, for example, that the parasitic invasion in *Shivers* functions as a cautionary tale regarding sexually transmitted diseases. The compelling AIDS parallel is simply chronologically impossible. A conscious suppression of political, as well as personal, references has allowed certain conceptual strategies to define his work.[26] This distancing effect is also evident in major strains of contemporary art that reject the mandates of identity politics or political correctness, however useful they may be as phases of cultural development, in favor of distinguishing art through its own terms. Up until this point, little of Cronenberg's direct autobiography is detectable in his films. He cites separation anxiety as his major childhood fear, noting *Bambi* and *Babar* – stories in which parents are killed leaving children alone – as being terrifying to him. But *The Brood* is unique in its uncanny resemblance to his personal life, and furthers his unapologetic approach to taboo subjects, in this case his monstrous depiction of motherhood. "It was also the beginning of an altogether darker phase of his work. The obsessions remained: medical science's misunderstood endeavors to assist the evolutionary process; the body's capacity to respond independently with transformation, mutation and its own creative diseases; descent into familial, societal or bodily chaos."[27]

EATON'S
Makes news on page 9
WEATHER
Wet, cold. High 7

THE TORONTO SUN

15¢
FINAL

VOL. 8 NO. 28 • 202,997 TORONTO, ONTARIO, WEDNESDAY DECEMBER 6, 1978 • 104 PAGES

BIZARRE SECOND KELLY MURDER

—REPORT, PICTURES SEE PAGE 3

Barton Kelly, of Halifax was found beaten to death last evening in Metro. Found near his body was the body of a grotesquely formed dwarf. Inside informents of the coroner's office report that the dwarf does not display human characteristics.

Fig. 18

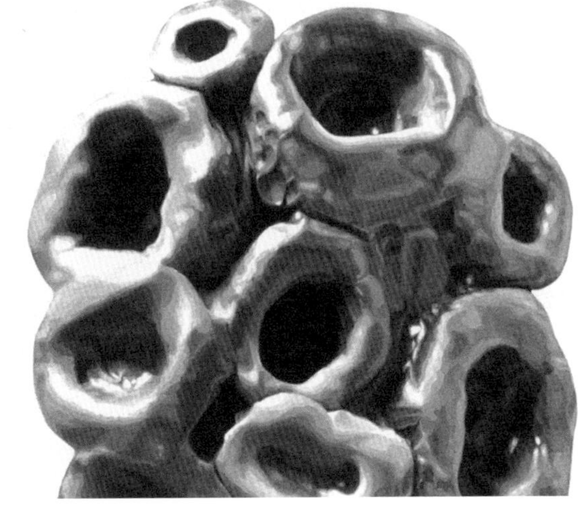

Fig. 19

While the increasing budgets, complexity of production, and investment in special effects promoted broader visibility for Cronenberg's films and refined a popular understanding of his work, his depiction of tyrannical organizational structures and their expression through corporate identities such as the Keloid Clinic (*Rabid*), Somafree Institute (*The Brood*), and Consec (*Scanners*), remains rooted in institutional architecture and decor. "With *Scanners* you're improvising the design of a whole invented world, suggesting a futuristic world without really saying it's futuristic. You're having to invent conference rooms and computer rooms, because you have to blow them up."[28] After *Scanners*, his most identifiably sci-fi film, he went on to make what many consider to be his major statement, *Videodrome*. This film "...represented the outer limits of Cronenberg's thematic concerns with wacky, well-meaning scientists and renegade, conspiratorial institutions: Professor Brian O'Blivian is the last of the director's absentee innovators, and Spectacular Optical the last menacing, covert operation."[29]

Fig. 20

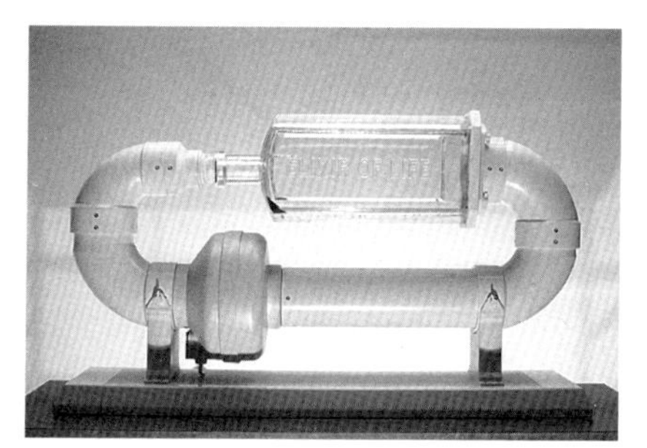

Fig. 21

It is from this decrepit eyewear store that the exhibition *Spectacular Optical* emanates. The front for Videodrome, a snuff tv program whose signal causes hallucination-inducing brain tumors, Spectacular Optical is the modest site of intended world domination: "We make inexpensive eyeglasses for the Third World, and missile guidance systems for NATO."[30] As a culminating institution, it incorporates a debased version of office décor complete with a super-graphic corporate identity that gives rise to a new technology that will forever alter the human condition. With *Videodrome*, Cronenberg presents a destabilized version of reality, with no cinematic distinction between the real and the hallucinatory. "The mediation of the image as a hyperlanguage and hyperreality...allows Cronenberg to situate his bodily figurations and demands a reading through the tropes of postmodernism, in which the negation of a mind/body dichotomy takes its place within a set of such negated oppositions and boundary dissolutions, including self/other, private/public, and spectacle/reality."[31] The exhibition pivots on similar thresholds characteristic of the postmodern moment, between rationality and its usurpation by a fetishistic unconscious, between doom and the euphoric embrace of "the New Flesh."[32] The false ending implied by the close of the century is also evoked by artists' tendency to resist conclusion, and to acknowledge the fragmented yet somehow evolving human condition.

In reference to his own clinical approach to "extreme forms of bodily damage," also evident in the work of contemporary artists, J.G. Ballard describes "the theatrical way that cadavers lie on their glass-top tables under the cool light of the dissection room. It's like a cross between a Conceptual art installation and a nightclub."[33] This outlook is not dissimilar to Cronenberg's, and it is a visual point of reference for the exhibition, *Spectacular Optical*. Strategies increasingly diffuse in contemporary artistic practice – such as the use of cinematic language or its breakdown into its constituent parts (set, prop, still, remnant); the mirroring of modern design, architecture, and décor; the acceptance of suburban ennui and its by-products as creative forces; or the positing of the mad scientist as artist – strategies that are designed to evoke cool terror, not unlike the evocations of Cronenberg's subversive temperament, are the locus of *Spectacular Optical*.

1. English subtitles for the lyrics of the final song from Masumune Shirow's animated film, *Black Magic M-66*: "From a printer that prints tomorrow, Blue data is overflowing."

2. As one critic notes: "In now-beginning premiere revivals, pic will be of interest to hardcore Cronenberg cultists for the appearance of numerous of the director's themes and fetishes in embryonic form." (Film Review of "Crimes of the Future," *Variety*, August 22, 1984): 17.

3. "...he has used his camera with a feeling for the tactile that is as evocative as the work of an Antonioni or a Warhol." (Review by Jacob Siskind "*Stereo* – an interesting first," *The Gazette*, Montréal, June 23, 1969); "In an academic garble, sounding like equal parts of the writings of Scottish psychiatrist R. D. Laing, and the narcotic ramblings of William Burroughs, the narration explains the project is the brain-child of Luther Stringfellow." (Peter Goddard, "*Stereo*," *The Telegram*, March 4, 1969); "Reminiscent by turns of Chris Marker's *La Jetee*, George Lucas' short USC version of *THX 1138* and Jean-Luc Godard's *Alphaville*, as well as 1960s New York underground scene, film abstractly examines the situation at the Canadian Academy for Erotic Inquiry, where eight individuals have been subjected to telepathic surgery." (Film Review of "*Stereo*," *Variety*, August 22, 1984): 17.

4. Jeff Wall, *Dan Graham's Kammerspiel* (Toronto: Art Metropole, 1991): 16-17. Cronenberg's fascination with the eccentricities of institutionalized power, particularly embodied by scientists who attempt to reinsert the erotic into science, also bears a strange parallel to the projects of Conceptual art: "The strategies of Graham, Buren, or Kosuth are, each in their own way, informed (through the issues raised by the institutionalization of Minimalism and Pop) by the combination of concepts drawn from the Frankfurt School tradition with related, historicist, critiques of urbanism. This combination took the form of linked studies in the development of state and scientific institutions (as mechanisms of social power and control) and research into the methods of siting these institutions within the modern city (or, more accurately, of the rebuilding of the modern city in terms of the strategic siting of these institutions). ...The influence of the Frankfurt School, but most particularly Walter Benjamin, is evident in the connections these authors make between their specific objects of study and the psychosocial effects of these objects, wherein mechanisms of power and domination are internalized by the atomized urban masses and involuntarily reproduced as profound estrangement." (Jeff Wall:13-15).

5. Peter Goddard, "Stereo," *The Telegram*, March 4, 1969.

6. Marshall Delaney, "Canada can't make big movies yet, but personal cinema is alive," *Saturday Night*, February 1970: 37.

7. Jennifer Taylor, "John Andrews: Architect," in Jennifer Taylor and John Andrews, eds., *John Andrews: Architecture a performing art* (New York: Oxford University Press, 1982): 17.

8. Scott Bukatman quotes Carrie Rickey in "Terminal Image," *Terminal Identity: the Virtual Subject in Post Modern Science Fiction* (Durham and London: Duke University Press, 1993): 82. "...Cronenberg is: 'a visionary architect of a chaotic biological tract where mind and body, ever fighting the Cartesian battle for integration, are so vulnerable *as to be easily annexed by technology*.'"

9. Scott Bukatman, "Terminal Image," *Terminal Identity: the Virtual Subject in Post Modern Science Fiction* (Durham and London: Duke University Press, 1993): 83-4.

10. As Jeff Wall describes the project: "Dan Graham's unrealized (and possibly unrealizable) project *Alteration to a Suburban House* (1978), generates a hallucinatory, almost Expressionist image by means of a historical critique of conceptual art. In this work, conceptualism is the discourse which fuses together three of the most resonant architectural tropes of this century (the glass skyscraper, the glass house and the suburban tract house) into a monumental expression of apocalypse and historical tragedy." (Jeff Wall: dustjacket)

11. Jeff Wall: 61.

12. Peter Goddard.

13. Discussion with Bryan Crockett, March 18, 1998.

14. Andrew Hultkrans, "Body Talk: Andrew Hultkrans talks with J.G. Ballard and David Cronenberg," *Artforum*, March 1997, XXXV, No. 7: 118.

15. Text from *Stereo*.

16. Chris Rodley, "Crimes of the Future: Obsessions and Avant-Garde Films," *Cronenberg on Cronenberg*, ed. Chris Rodley (London: Faber and Faber, 1992): 26.

17. Chris Rodley, "Unrequited Life: From Dead Zones to Dead Ringers," *Cronenberg on Cronenberg*, ed. Chris Rodley (London: Faber and Faber, 1992): 145-46.

18. Atom Egoyan, "Atom Egoyan Interviews David Cronenberg," *Take One*, Film in Canada, No. 3, Fall 1993: 11.

19. Chris Rodley, "Introduction," *Cronenberg on Cronenberg*, ed. Chris Rodley (London: Faber and Faber, 1992): xviii.

20. Chris Rodley, "Body Talk: The Cinepix Years," *Cronenberg on Cronenberg*, ed. Chris Rodley (London: Faber and Faber, 1992): 58.

21. Jeff Wall: 42.

22. Chris Rodley, "Body Talk: The Cinepix Years," *Cronenberg on Cronenberg*, ed. Chris Rodley (London: Faber and Faber, 1992): 54.

23. Chris Rodley, "New Flesh for Old: The Tax-Shelter Experiments," *Cronenberg on Cronenberg*, ed. Chris Rodley (London: Faber and Faber, 1992): 75.

24. Text from *Videodrome*.

25. Chris Rodley, "New Flesh for Old: The Tax-Shelter Experiments," *Cronenberg on Cronenberg*, ed. Chris Rodley (London: Faber and Faber, 1992):99.

26. Conversation with Elisabeth Subrin, March 1998.

27. Chris Rodley, "Introduction," *Cronenberg on Cronenberg*, ed. Chris Rodley (London: Faber and Faber, 1992): xviii.

28. Chris Rodley, "New Flesh for Old: The Tax-Shelter Experiments," *Cronenberg on Cronenberg*, ed. Chris Rodley (London: Faber and Faber, 1992): 86.

29. Chris Rodley, "Introduction," *Cronenberg on Cronenberg*, ed. Chris Rodley (London: Faber and Faber, 1992): xix.

30. Description of Spectacular Optical from *Videodrome*.

31. Scott Bukatman, "Terminal Image," *Terminal Identity: the Virtual Subject in Post Modern Science Fiction* (Durham and London: Duke University Press, 1993): 85.

32. Regarding "the New Flesh" it is interesting to note that: "One alternative ending [to *Videodrome*] was a mutated transsexual orgy in the Videodrome chamber. After Max shoots himself (the last image of the final version), we might have seen Bianca O'Blivion (Sonja Smits), Max (James Woods), and Nicki Brand (Debbie Harry) sexually entwined, all in each other. 'A happy ending? Well, my version of a happy ending – Boy meets Girl, with a clay wall maybe covered with blood. Freudian rebirth imagery, pure and simple.' Max's imagined abdominal vagina was here to be matched by Nicki's and Bianca's newly found penises (á la *Rabid*). Male and female mutated sex-organ appliances were designed, but Cronenberg decided to drop the final scene altogether. Constant references in *Videodrome* to 'the New Flesh' may have been clarified by this vision: another, more inventive, satisfying fleshy existence waiting just on the other side of death. [...] Max shooting himself was the right ending for the movie. And it's almost the same ending as *Dead Ringers, The Fly* and *The Dead Zone*. On each of those films there was a coda written that never ended up in the picture." (Chris Rodley, "New Flesh for Old: the Tax-Shelter Experiments," *Cronenberg on Cronenberg*, ed. Chris Rodley (London: Faber and Faber, 1992): 97.)

33. Andrew Hultkrans, "Body Talk: Andrew Hultkrans talks with J.G. Ballard and David Cronenberg," *Artforum*, March 1997, XXXV, No.7: 78.

Fig. 1
Jeremy Blake, *The Cold Room*, 1998, digital c-print; courtesy of the artist and Bronwyn Keenan, New York

Fig. 2
Invitation to the screening preview of *Stereo*, c. 1969, offset on card, Collection of David Cronenberg Papers, Cinematheque Ontario, Film Reference Library, Toronto; photo: Thomas Moore

Fig. 3
John Andrews, Scarborough College, University of Toronto, 1963; photo: David Moore

Fig. 4
David Cronenberg shooting *Stereo*, with actor Ronald Mlodzik, c.1969; courtesy David Cronenberg Productions, Toronto

Fig. 5
Luisa Lambri, *"Plan Libre (Out of Site)"*, 1997, cibachrome print; courtesy of the artist

Fig. 6
Still from *Crimes of the Future*, c.1970, photo: New Cinema Canada

Fig. 7
Charles Long, *Untitled*, 1993, plastic, steel; private collection

Fig. 8
Bryan Crockett, *Untitled*, 1997, latex balloons, epoxy resin; courtesy of the artist

Fig. 9
Shellburne Thurber, *Motel Room with Dropped Ceiling*, 1990, ektacolor print; courtesy of the artist and Jack Shainman, New York

Fig. 10
Randall Peacock, *room tone*, 1998, work in progress; courtesy of the artist

Fig. 11
The Parasite Murders, 1975, movie poster, offset on paper; Collection of David Cronenberg Papers, Cinematheque Ontario, Film Reference Library, Toronto; photo: Thomas Moore

Fig. 12
Jason Fox, *Dreven*, 1994, acrylic on canvas; courtesy of the artist and Feature Inc. New York

Fig. 13
Bill Stonehan, Drawings of arm and hand movements of the Fly Creature, c.1985; Collection of Chris Walas Inc.; photo: Ben Blackwell

Fig. 14
Bonnie Collura, *Model for Nymph*, 1997, ink on paper; Collection of Kenneth L. Freed; courtesy Basilico Fine Arts, New York

Fig. 15
Scene from *Rabid*, 1976: photo: Cinepix, Inc.; courtesy of Cinematheque Ontario, Toronto

Fig. 16
David Cronenberg shooting *Shivers*, 1975; courtesy of Cinematheque Ontario, Toronto

Fig. 17
Scene from *Videodrome*, 1982; photo: Universal City Studios Inc.; courtesy of Cinematheque Ontario, Toronto

Fig. 18
Newspaper "*The Sun, Dec. 6, 1978*" used as prop in the film, *The Brood*, c. 1980, newsprint; Collection of Film Reference Library, Cinematheque Ontario/Festival of Festivals, Toronto; photo: Thomas Moore

Fig. 19
Alex Ross, *Untitled*, 1997, oil on canvas, photo: Oren Slor; Collection of Mark Safan; courtesy Feature Inc., New York

Fig. 20
Mariko Mori, *Mirage*, 1997, digital video still formatted on glass panel; courtesy Deitch Projects, New York

Fig. 21
Julian LaVerdiere, *Wind Tunnel Model*, 1995; courtesy of the artist

Fig. 22
Still from *Videodrome*, 1982; Universal City Studios Inc.

Fig. 22

David Cronenberg has commented, on his decision to make the film *Crash*: "Eventually, after procrastinating (in the way that writers do in writing the script), I started to feel that all the necessary connections – those strange, mysterious tissues that bind you to something – were healthy and working." This could be said of any project that slowly infects the imagination, and certainly it is true of *Spectacular Optical.*

I would first like to thank all of the participating artists and writers, whose talents have made this exhibition and publication possible. We are indebted to the following collectors and institutions for the loan of artworks in the exhibit: Bill Arning; Luhring Augustine; Basilico Fine Arts; David Cronenberg Papers, Cinematheque Ontario, Film Reference Library, Toronto; Galerie René Blouin, Montréal; Feature Inc.; Gallerie Emi Fontana; Kenneth L. Freed; Sandra Gering; Jay Gorney Modern Art; Pat Hearn Gallery; Bronwyn Keenan; Lisson Gallery, London; Steffany Martz; Metro Pictures; Museum of Modern Art of Rio De Janeiro, Brazil; Mark Safan; Jack Shainman; Carol Spier & James McAteer; and anonymous lenders.

Spectacular Optical was initiated by Gregory Crewdson, Sandra Antelo-Suarez, and Thread Waxing Space Founder and Trustee Tim Nye. *Spectacular Optical* is the debut exhibition of curator Lia Gangitano, who took the idea for this exhibition and shaped it with her singular vision. She was aided in her efforts by Cronenberg's archivist Fern Bayer and Associate Producer Sandra Tucker, whose involvement was critical to the success of this exhibition. Lia would also like to thank: Jacqueline Humphries for her editing expertise and friendship; Leo Villareal for his encouragement and interest in this project; Rosemary Ullyot and Sylvia Frank at The Film Reference Library for generously providing access to their invaluable archive; Jeremy Blake for recommending vital source material for her essay; intern Aprile Age for her research assistance; Hudson, Stefano Basilico, Rupert Goldsworthy, Todd Levin, Ihor Holubizky, Joe Wolin, Liz Kotz and Milena Kalinovska for their insight and advice; Jack Fahey, Tim Obetz and Tim Roche at Haverford Systems for their technical support.

The success of any exhibition at Thread Waxing Space is due to the unwavering support from the Board of Trustees. They are: Timothy U. Nye, Miguel Abreu, Adam Ames, Jedidiah Alpert, Dr. Fredric S. Brandt, Leonardo Drew, Jacqueline Humphries, Joy Mountford, Robert Reynolds, Carol A. Sawdye and Leopoldo Villareal. Thread Waxing Space's staff deserves special mention for their tireless efforts and enthusiasm. *Spectacular Optical* is a major endeavor and Vanessa Baish, David Freilach, Lia Gangitano, Eric Manigian and C. Jo Whitsell made it possible. Special thanks to Corey McCorkle for his design support. Under the expert direction of Eric Manigian, thanks go to installers Mario Acosta, Bill Feeney, Henry Levine, Vicki Pierre, Rafael Sanchez and Rob Sinclair.

This publication would not have been possible without the motivation and conviction of Sandra Antelo-Suarez and Michael Mark Madore of Passim. Thanks also to Passim Assistant Editor Greg Gangemi for his effort and commitment. Klaus Kempenaars and Pascale Willi of x-height are responsible for realizing the elegant design of this publication.

I would like to express my gratitude to Bonnie Clearwater of MoCA, North Miami and Susan Krane of UC, Boulder for their early commitment to presenting this exhibition at their esteemed institutions.

Exhibition support has been generously provided by The Peter Norton Family Foundation, the Canadian Consulate General-New York, presentation Funds of the Experimental Television Center, Haverford Systems, and public finds from the New York State Council on the Arts,a State Agency.

Finally, I would be remiss if I did not take this opportunity to recognize the enormous contribution of David Cronenberg, whose uncanny vision and point of view are at the center of this discussion.

Ellen F. Salpeter, Executive Director
Thread Waxing Space

This publication accompanies the exhibition *Spectacular Optical*, curated by Lia Gangitano and organized by Thread Waxing Space, N.Y.

Spectacular Optical has been made possible by the generous donations of: Jimmy Belilty, The Peter Norton Family Foundation, the Canadian Consulate General – New York, Presentation Funds of the Experimental Television Center, Haverford Systems, and public funds from the New York State Council on the Arts, a State Agency.

Publisher of TRANS >arts.cultures.media
109 West 17th Street
New York, NY 10011

Co-published by TRANS>arts.cultures.media
and Thread Waxing Space, N.Y.
ISBN: 1-888209-04-6
Library of Congress Catalog Card Number: 98-67113

EDITORS
Sandra Antelo-Suarez and Michael Mark Madore

ASSISTANT EDITOR
Gregory Gangemi

DESIGN
x-height, Klaus Kempenaars and Pascale Willi

PRODUCTION
Becotte & Gershwin, Inc., Warminster PA

EXHIBITION TOUR

Thread Waxing Space, N.Y.
May 28, 1998 – July 18, 1998

Museum of Contemporary Art, Joan Lehman Building,
North Miami, FL
October 18, 1998 – November 29, 1998

CU Arts Galleries, University of Colorado at Boulder
Fall 1999

CREDITS

Kathy Acker, from *Bodies of Work*, reprinted by permission of Serpent's Tail, London.

Jean Baudrillard, "The Xerox Degree of Violence" first appeared in French in the book titled *Ecran Total*, Galilée, 1997. It has been translated into English for the first time for this book, by permission and under the supervision of the author.

Scott Bukatman, "Videodrome," from *Terminal Identity*, reprinted by permission of Duke University Press, © 1993.

Judith Butler, from *Gender Trouble*, reprinted by permission of Routledge Press, Inc., © 1990.

Slavoj Žižek, "The Spectralization of the fetish," "The Suspension of the Master," and "Moving statues, frozen bodies" from *The Plague of Fantasies*, first published by Verso, 1997 © Slavoj Žižek 1997, reprinted by permission of the author.